ENGINEERING TOOLS, TECHNIQUES AND TABLES

MANUFACTURING ENGINEERING

ENGINEERING TOOLS, TECHNIQUES AND TABLES

Additional books in this series can be found on Nova's website under the Series tab.

Additional E-books in this series can be found on Nova's website under the E-books tab.

ENGINEERING TOOLS, TECHNIQUES AND TABLES

MANUFACTURING ENGINEERING

ANTHONY B. SAVARESE
EDITOR

Nova Science Publishers, Inc.
New York

For permission to use material from this book please contact us:
Telephone 631-231-7269; Fax 631-231-8175
Web Site: http://www.novapublishers.com

Additional color graphics may be available in the e-book version of this book.

Library of Congress Cataloging-in-Publication Data

Manufacturing engineering / editor, Anthony B. Savarese.
p. cm.
Includes bibliographical references and index.
ISBN 978-1-61209-987-3 (hardcover)
1. Manufacturing processes. I. Savarese, Anthony B.
TS183.M345 2011
670.42--dc22

2011007088

Published by Nova Science Publishers, Inc. † New York

CONTENTS

PREFACE

This new book presents current research in the study of manufacturing engineering, including the theory and practice of virtual manufacturing; adapting local economies to market changes; conveyor systems balancing; optimization of manufacturing conditions of brake friction materials; the impact of sanctions on the buyer-supplier relationship; processing of electrolytic capacitors used widely in electrical circuits and recent developments in automated manufacturing systems.

Chapter 1 - This chapter gives a general description of simulation and its associated system of interest. In the context of virtual manufacturing, three domains can be distinguished; product domain, process domain and resource domain. Examples of simulation in these there domains are given, as well as some examples of simulation across these domains. Typical steps/phases in a simulation project are described, as well as common pitfalls. In industrial simulation projects, usually a number of stakeholders are involved with different maturity/experience in the field of simulation. It is described how such industrial simulation projects can be supported by a handbook, developed in close collaboration with a group of companies. As one example of advanced applications, simulation-based remote monitoring and diagnostics is described. The other example of advanced applications given in the paper is that of simulation-based optimisation. Many simulation tools and projects aim at providing decision support to a human decision maker. High level information fusion, a development originating from defence research, also aims at providing decisions support. A comparison between virtual manufacturing and information fusion reveals that a popular reference model for information fusion called JDL-model is very apt to serve as a reference model for virtual manufacturing.

Chapter 2 - Today's economic downturn is providing decision makers with an understanding of the importance to dramatically speed up the capability to adapt local economies to market changes. To align production and consumption patterns to sustainable performances at local and global level is a priority for both public and private actors. According to that, since adaptation strategies encompass multiple interactions between individual and collective actions, the complexity of the challenge lies in how to coordinate and implement *micro-* and *macro-*level theories in practice.

How to benefit from synergies between regional transformations and corporate reorganization? How to make these synergies effective? For trying to answer these questions entails we developed an in-field investigation on how to invest in inter-public-private partnerships as well as in inter-organizational linkages.

Our time is characterized by accessibility and comprehensiveness of information and, in turn, by its effects on society. Social networks represent a new

Chapter 3 - This chapter focuses on two special queueing networks composed of a decision maker (coordinator) and K CSPSs (Conveyor-Serviced Production Stations), originaled in Matsui, 1982. One is a series' array of CSPSs (Model I), and the other is an ordered-entry array of CSPSs (Model II). The chapter first presents a station-centered approach to the class of CSPS Network, and prepares a few queueing formulas and general relation concerning average criteria. Next, a 2-level mathematical formulation of max-max and max-min types is explained and presented for determining both the cycle time and time-range (control variables) of optimizing the production rate of Models I and II ,respectively. Through the chapter, a mathematical theory (or method) concerning CSPS Network is systematically developed and a typical case and numerical consideration is given under regular or Poisson arrival and general service with equal mean. Finally, it is remarked that this original material is already written in Matsui (1982), and is recently developed to the cost factor case in Matsui (2008).

Chapter 4 - Process of brake friction material manufacturing has a crucial impact on the future properties of brake friction material. It is especially related to the level and stability of friction and wear during braking. Development of brake friction material is strongly affected by its formulation and manufacturing conditions. Due to complex and interrelated influences between formulation and manufacturing conditions of brake friction material, it is difficult to find out the best set of manufacturing conditions, for the specific formulation of friction material, which satisfy wanted friction and wear properties. In this chapter, influence of manufacturing conditions of the brake friction material has been investigated versus its wear. In order to investigate the character of influence of manufacturing conditions, effect of manufacturing parameters on wear properties of the brake friction material has been modelled. This model was able to recognize the way of how moulding pressure, moulding time, moulding temperatures, heat treatment time, and heat treatment temperatures affecting wear of the brake friction material. The model of influence of manufacturing parameters on wear of the brake friction material was based on artificial intelligence. The model is able to make a functional relationship between a formulation of friction material and effects of chosen manufacturing parameters on its wear. It provided possibilities for searching the best set of manufacturing conditions and their adaptation to a formulation of brake friction material.

Chapter 5 - Buyer-supplier relationships have received increasing research attention in the last two decades. Empirical evidence shows that relationship quality can be examined by considering a number of relationship factors. Literature also shows that there are time-based features of relationship quality. Reviewed literature pinpoints a number of research gaps in this dyadic relationship context, one of which is the time-based feature context. Among others, empirical research gaps in this domain also include the geographical context.

Combining the geographical and time-based context research gaps, a pilot study explored buyer-supplier relationships between Libyan buyers and foreign suppliers over time (Tantoush et al., 2009). The Libyan economy and in particular the Libyan oil industry was adversely affected by the imposition of sanctions by the US and the UN from 1986 to 2003. Libya's refining sector was hardest hit by UN Resolution 883 of 11 November 1993, banning Libya from importing refinery equipment. At the time, Libya was seeking a comprehensive upgrade to its entire refining system, with a particular aim of increasing output of gasoline and other light products (e.g. jet fuel).

The pilot study shapes the needs for further investigation in this area, and a proper research framework is needed in order to collect empirical data accordingly. This chapter aims to provide a commentary of the pilot study, and then guidelines for further research. More importantly, the reflections from this pilot study will be shared. The chapter will first briefly summarise the key findings of the pilot study. It will then present the implications and the recommendations for further study. After that, the benefits for conducting this pilot will be investigated.

Chapter 6 - The electrolytic capacitor is an electronic device of great technological importance because it is widely used in electrical circuits. Electrolytic capacitors are used in electronic equipments of defense, information technology, industrial communication, tools for office, automation, robotics, etc. The preference for electrolytic capacitor is related to its high specific capacitance (high value of capacitance in small volume). Tantalum is the main material used for manufacturing of electrolytic capacitors of high performance. Only its high costs restrict a wider application. This fact has stimulated scientific research in order to establish a lower cost substitute material. This chapter focuses on the production and characterization of the alloy Nb46wt.%Ta powder. The substitution of pure tantalum by the alloy Nb46wt.%Ta results in economical benefits. The reasons are: i) the substitution of tantalum by niobium by 46 wt.% results in a cheaper material for electrolytic capacitors; and ii) powder production of the alloy Nb46wt.%Ta using aluminothermic step and subsequent comminution operation is cheaper.

Chapter 7 - This chapter is to describe the recent development and applications of automated and high speed manufacturing technology in industrial production. It includes the general application of programmable logic control, studies and design of automated and high speed product assembly line, computer aided design of automated manufacturing systems, computer aided manufacturing simulation, and future trend of automated manufacturing technology. Several case studies in this chapter aim at the introduction, study and analysis for automated and high speed manufacturing and production. The application of programmable logic control to industry brings revolution for the manufacturing techniques. It allows more sophisticated, flexible, reliable, and cost-effective manufacturing process control. Automation is to use control system to reduce human labor intervention during manufacturing processes and production. It plays very important role and puts strong impact in today's industries. Automation is not only significantly increasing the production speed but also more accurately controlling product quality. The automated manufacturing can maintain consistent quality, shorten lead time, simplify material handling, optimize work flow, and meet the product demand for flexibility and convertibility in production. Computer aided engineering design can quickly model the automated manufacturing systems and speed design and development cycle. Computer aided manufacturing can improve engineering integral processes of product design, development, engineering analysis, and production. The current economic globalization requires significant labor cost reduction through industry automation, improved machine tools, and efficient production process.

Based on author's current research projects, several case studies in fully automated and high speed manufacturing systems have been introduced and analyzed in this chapter including fully automated and high speed assembling processes of gas charging system, high viscous liquid filling system, and ultrasonic welding of cap system. All these fully automated and high speed manufacturing systems developed by author have been analyzed and verified through preliminary prototypes or field tests. The computer aided simulation and testing results indicated the reliable performance, feasible function, cost-effective mechanism, increased productivity, and improved product quality by these fully automated and high speed manufacturing systems.

In: Manufacturing Engineering
Editors: Anthony B. Savarese
ISBN: 978-1-61209-987-3

Chapter 1

VIRTUAL MANUFACTURING THEORY AND PRACTICE

Leo J. De Vin
Virtual Systems Research Centre – Intelligent Automation,
University of Skövde, Skövde, Sweden

ABSTRACT

This chapter gives a general description of simulation and its associated system of interest. In the context of virtual manufacturing, three domains can be distinguished; product domain, process domain and resource domain. Examples of simulation in these there domains are given, as well as some examples of simulation across these domains. Typical steps/phases in a simulation project are described, as well as common pitfalls. In industrial simulation projects, usually a number of stakeholders are involved with different maturity/experience in the field of simulation. It is described how such industrial simulation projects can be supported by a handbook, developed in close collaboration with a group of companies. As one example of advanced applications, simulation-based remote monitoring and diagnostics is described. The other example of advanced applications given in the paper is that of simulation-based optimisation. Many simulation tools and projects aim at providing decision support to a human decision maker. High level information fusion, a development originating from defence research, also aims at providing decisions support. A comparison between virtual manufacturing and information fusion reveals that a popular reference model for information fusion called JDL-model is very apt to serve as a reference model for virtual manufacturing.

1. INTRODUCTION TO "SIMULATION"

There are many different definitions of "simulation", some related to computer simulation, and others more general. What these definitions tend to have in common is that a model is used to depict a "system of interest" or "situation/scenario of interest". This model can be a computer model, but it might for instance just as well be a physical model or mental model. In social psychology for instance, participants in experiments can be confronted with a certain situation or scenario, and are then asked how they would respond if this were a real situation. Children's play is another interesting example of non-technical simulation. Imagine for instance a "cowboys and indians" game for which one of the children suggests the use of a wireless radio set. Not unexpectedly, the use of this radio set will be an issue of discussion. What these children essentially do is "VV&A": verification (is the use of a radio set correct, given the context of the game they intend to play), validation (would it affect the game) and acceptance. Sometimes, the group may allow its use and sometimes not, depending on their mood (do they want to emulate "cowboys and indians" accurately or does it not matter for the fun of playing?). In this example, the purpose of the game plays a central role, just like for industrial simulation projects, where the decision "accept" or "reject" of a model can depend heavily on the intended purpose.

Within the context of this chapter, we can describe simulation as "experimentation with some model of a system of interest (SoI)". This SoI may be an existing system, a projected system, or a completely imaginary system. As mentioned before, the term SoI may also stand for Situation/Scenario of Interest. Simulation helps us to reduce uncertainty in the decision making process by enabling us to explore many alternatives against relatively low costs and by presenting knowledge about projected artifacts on various levels of abstraction. Simulation as a communication tool also enables us to obtain a refined specification of customer requirements by exposing prospective customers to alternative design solutions in a virtual environment. A prominent question in any simulation is how much we "trust" the simulation results. This role of simulation in the innovative industrial process also makes us more aware of the lack of a suitable science base for engineering work. At best, people try to apply modified theory of science to engineering. This, however, results in an obvious mismatch as science deals with asking questions about the existing whereas engineering is about creating the (presently) non-existing [1].

The relationship between the System of Interest and its associated model is shown in Figure 1. The model is an abstract representation of the SoI and ideally, from the behaviour of the model, conclusions can be drawn concerning the SoI. Likewise, from the observed behaviour of the SoI, conclusions can be drawn concerning the model.

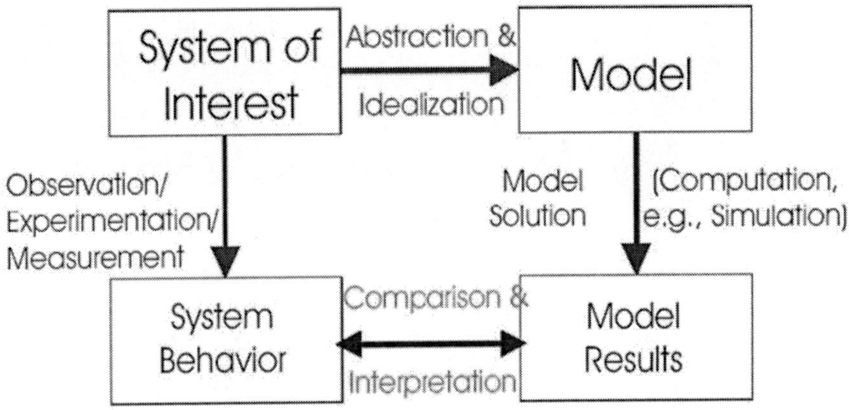

Figure 1. System of Interest and Model [2].

In a working paper of the Virtual Systems Research Centre at the University of Skövde, five main different uses of simulation and virtual systems are distinguished, including:

Simulation as a tool to study an SoI in order to create new knowledge about the modelled system, or to refine existing knowledge about it. This new or improved knowledge can subsequently be used for decision support. An important application of this use of simulation is “what-if” analysis.

Simulation as a tool to train operators in the use of the SoI. In this case, the simulation model serves as a means to transfer knowledge about the SoI to the operators.

Serious gaming as a way to create situations that are realistic, even although the situation itself may never occur. Serious gaming is often used to train people and organisations for situations in which communication, coordination and decision making are important, such as in complicated rescue actions or natural disasters. In manufacturing engineering, it can be used to train aspects such as lean production or production planning in general.

Simulation as a way to test and benchmark for instance production planning algorithms. Testing and comparing algorithms or other solutions (such as soft computing) in a real production environment is not practically possible, but it is possible to do so in simulated environments.

Simulating a situation and/or sequence of events in order to test peoples' attitudes or responses. This includes responses of people to (prototypes of) new products.

Gaming primarily as entertainment. Nevertheless, such games can include various educational moments.

2. Examples of Manufacturing Simulation Across PPR Domains

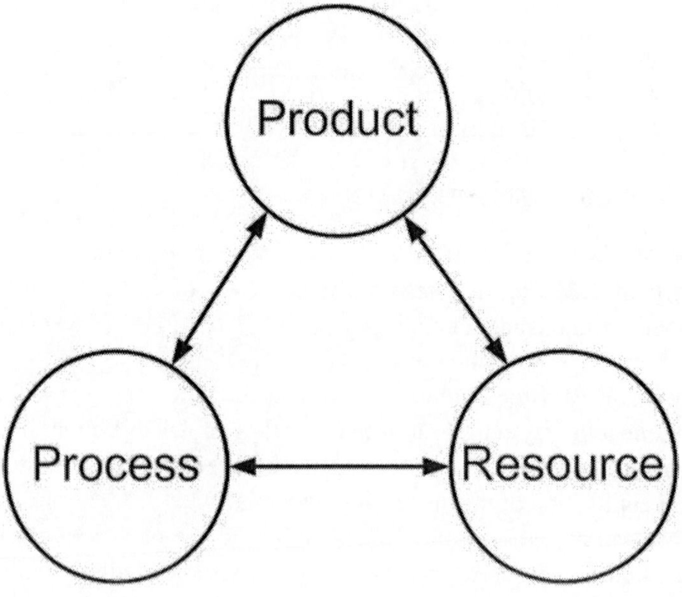

Figure 2. The PPR hub.

Within the context of an integrated approach to product- and production development, it is mandatory to consider products, processes and resources

("PPR", also called "PPR hub", see Figure 2) concurrently. This division into three domains is also very suitable when discussing different types of simulation or applications of the same modelling & simulation technique in different domains. It also indicates that there may be many orders of magnitude in time scales or spatial scales, for instance between simulation-in-the-loop for in-process control of sheet welding processes and discrete event simulation of the product flow in a production plant, an issue also highlighted by the NSF Blue Ribbon Panel on Simulation-Based Engineering Science [3].

2.1. Modelling and Simulation in the Product Domain

Many aspects of products may be simulated, but for sheet metal products, mainly physical properties are of interest. Two well-known methods that are also commonly used in industry are the finite element (FE) method and multi-body system (MBS) simulation. The importance of simulation in product development in general is described in [4]. It facilitates the sliding between existing and not yet existing objects, and supports the designer in decision making through the possibility to explore "what if" scenarios.

The FE method is a general method to model and simulate the physical behaviour of bodies with arbitrary shape. FE simulations have mainly be used as a tool for detailing components, but there is a need to use the FE method in the earlier phases of the design process. There is a trend to qualify detailed behaviour of complex artefacts with FE simulations assisted by reduced testing, for instance [5]. The computer resources required for an FE simulation grows exponentially with the size of the model; the FE method is thus not directly scalable. This problem is addressed by several modelling methods that are variations on the approach to synthesize models of complex technical systems from condensed FE submodels.

Modelling of the dynamic behaviour of a multi-body system is characterised by a composition of rigid bodies, interconnected by joints, springs, dampers, and actuators. Force elements such as springs, dampers acting at discrete attachment points result in applied forces and torques on rigid bodies. Joints constrain the motion of the bodies in the system. However, especially for sheet metal components, the rigid body assumption is in many cases a too crude approximation. In order to remedy this, many MBS software packages such as ADAMS have capabilities to import condensed FE models. There are several complementary condensation methods available for generating a reduced FE problem. The most widely known are static condensation and component mode synthesis [6].

2.2. Modelling and Simulation in the Process Domain

In the process domain, the manufacturing processes are modelled and simulated. Examples of processes are punching, laser cutting, welding, and a variety of bending/forming processes such as air bending, deepdrawing, hydroforming, and laser bending. For 3-D forming process, mainly FE modelling and simulation is used, for instance [7, 8]. Modelling and simulation of brakeforming and more in particular air bending has been the focus of renewed attention during the nineties, but reasonably recent articles [9, 10] and recent industrial projects [11] show that modelling and simulation of these processes still receives ample attention in academia and industry. Models for air bending are often FE models [12, 13] or analytical models [14, 15]. Examples of the application of process models for in-process control can be found in [16, 17]. FE models can be used for studies of sheet metal working processes [18] as well as for process related studies such tool and press brake deflections in bending [19] or tool performance for punching [20].

2.3. Modelling and Simulation in the Resource Domain

In the context of this chapter, modelling & simulation in the resource domain is restricted to 3-D graphical simulation of manufacturing machinery (such as robots) and production lines. The simulation technologies considered are Discrete Event Simulation (DES, sometimes also referred to as "production flow simulation", Figure 3) and Computer Aided Robotics (CAR, sometimes also referred to as "geometry simulation" or "continuous simulation", Figure 4). DES tools are normally used to study the product flow in a manufacturing facility, although their use can be extended to other areas than manufacturing such as healthcare [21]. CAR tools are normally used to study movements of industrial robots and so on, with collision avoidance and off-line programming as important applications. The CAR model of the cell in Figure 4 shows a press with a materials handling robot that feeds the press with sheets from the de-stacker and handles stamped parts and waste material by placing these onto two different conveyors. A technology related to CAR is ergonomic simulation, which is used for instance for posture analysis in workplace design [22].

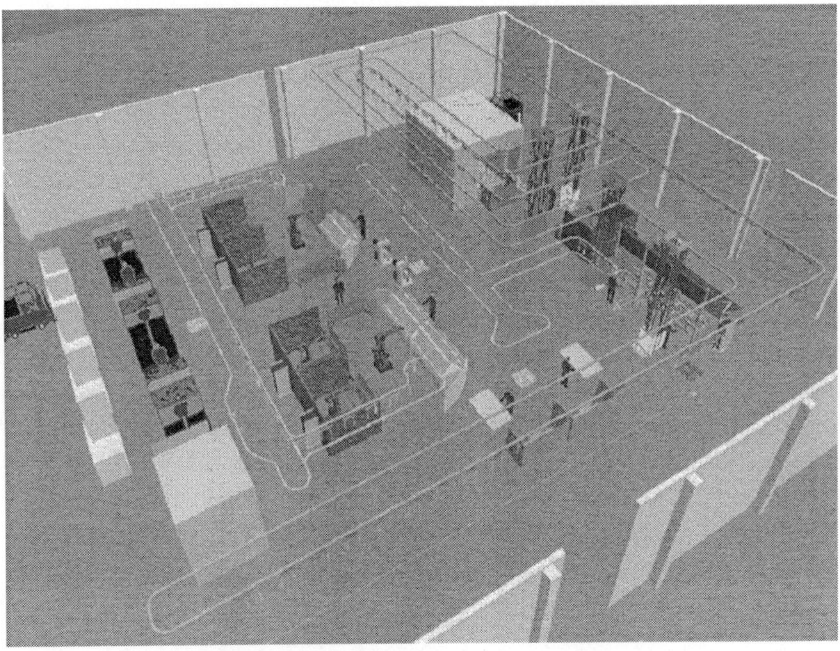

Figure 3. Animation snapshot from DES.

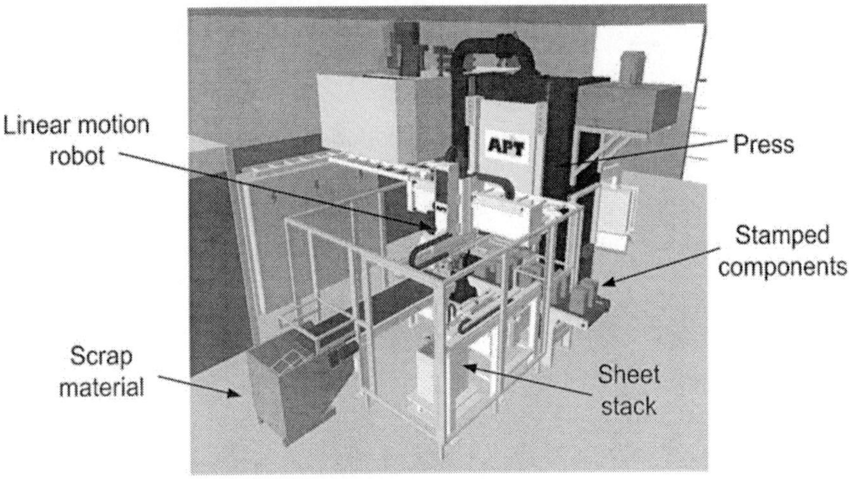

Figure 4. Animation snapshot from CAR.

2.4. Co-Simulation across PPR Domains

Co-simulation is the integrated use of different types of simulation tools to solve a problem. One such example is the use of condensed FE models in MBS simulation as discussed above. Whilst co-simulation across PPR domains has a large potential as a support for concurrent engineering, examples of co-simulation across different PPR domains are still scarce. One example is the integration of a process model into a CAR tool for the simulation of a robot assisted press brake (with the robot supporting the product during bending) [23]. Another example is the integration of sensor simulation with CAR [24]. This application is used for process in the loop simulation of sheet metal welding with actual sensor data, i.e. the sensor data of the welding process is input for the simulation model which in turn is connected to the robot controller. This allows for a tight integration between physical world and virtual world, and is used for seam tracking and obstacle avoidance. An additional advantage is that the simulation model can "look ahead" which helps to avoid robot poses that represent (or are near) singular points.

3. Simulation Projects

3.1. Phases an a Simulation Project

A number of activities can be distinguished in a simulation project. Most authors present these activities in the form of a flow chart. However, such flowcharts should not be treated too rigidly, as a lot of zigzagging between activities/phases may take place. Furthermore, the availability of data for instance may influence the definition of the goal. An example of such a flowchart is given in Figure 5 [25]. Short descriptions of the main activities/phases are given below.

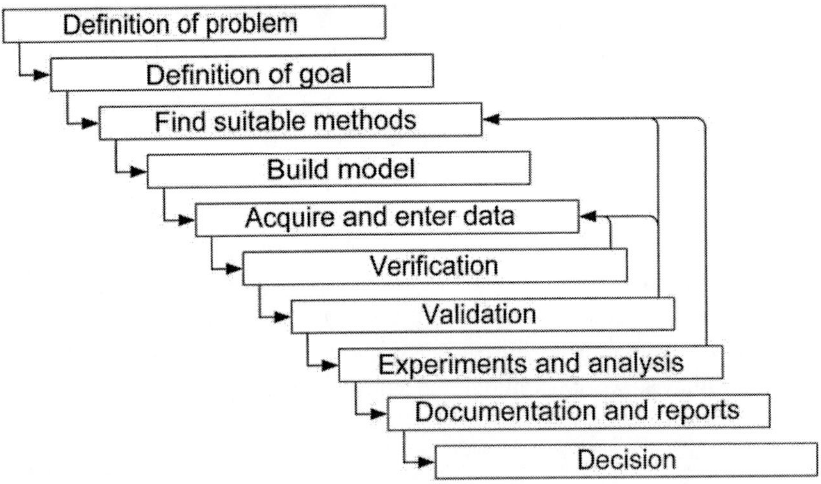

Figure 5. Activities in a simulation project (after [26] and similar decompositions, for instance [27]).

3.2. Definition of the Goal

When embarking on a simulation project, there are some important things to consider. First of all, a simulation is normally carried out to get an answer to certain questions, or to support certain decisions. Therefore, it is very important to design the simulation project in such a way that it will be able to answer the questions at hand. Therefore, the problem/question (or decision to be supported) must be defined with great care. However, before embarking on a simulation project, it can be very worthwhile to consider other solutions. Simulation is no goal in itself, and a straightforward analytical solution may be easier to arrive at. Furthermore, if no detailed answer is needed, then an upper bound / lower bound analysis (or best case / worst case) may provide a sufficiently accurate answer. Or, in some cases, a simple paper-based simulation can give the answer, for instance by using a Gantt chart for production planning problems. The goal of the project may be influenced by a number of issues. As mentioned, the goal normally is to answer a certain question, but if there are problems with for instance data-acquisition, then the goal must be adjusted to what is practically possible. Furthermore, if the goal is too ambitious for the resources allocated, then there is a large risk that the project will be completed late.

3.3. Building the Model

When building a simulation model, one has to keep the purpose of the simulation model in mind. Models need to serve their purpose, and a crude model that is completed in time but does the job is better than a detailed model that is completed late. However, if the purpose includes (or may include) extended use of the model, then this extended use needs to be taken into account. This usually means that it initially takes longer to build the model, but this saves time in the end [4]. Extended use may mean that the model needs to be extendable, refinable, and/or that it initially serves as a stand-alone model but in later phases becomes integrated with business process and information systems.

3.4. Data-Acquisition

Data-acquisition for a simulation is a crucial task and not seldom, problems arise during this phase. There could, for instance, be too little data or too much, data can be unreliable, incomplete or inconsistent. Other potential problems include multiple entries of the same data. In automated manufacturing, often many types of data are collected, but the meaning of this raw historical data is not always clear enough to use it for simulations or decision making. Historical data is not always representative for the situation to be simulated as typically, companies carry out production improvement processes, change maintenance policies, and so on. Manual data collection by operators has the disadvantage that small stops often are not reported. Manual data collection by an observer has the disadvantage in manual labour that this typically affects the performance in a positive or negative way. There are even examples from mixed automated/manual labour in which workers slowed down machines by reducing the feed-rate to make the manual labour look more efficient.

3.5. Verification and Validation

It is good practice to have an attitude like: "The results from a simulation are to be considered as being inherently inaccurate, unless there are good reasons to believe the opposite". In other words, one needs to be able to motivate why the simulation produces realistic results. Important steps are verification (checking the model for errors) and validation (checking whether the model produces realistic overall results). Unfortunately, model building and data acquisition often take more time than planned, and in many cases the simulation engineer is very keen to start with simulation runs (either through fondness with simulation or by pressure from project management). Rather often, this goes at the cost of proper verification and validation. An overall validation at the end is not sufficient, as the model may contain certain errors that compensate each other under certain circumstances. Such errors need to be detected during verification.

It is good practice, like in software development and testing, to separate the model development and the verification/validation. This separation is good for two reasons. Firstly, developers tend to test things that required extra attention during development and sometimes forget to test "trivial" things. Secondly, a separate tester is more objective due to the absence of an "emotional bond" with the model. Unfortunately, in practice it is rather often the simulation engineer who carries out the verification and validation.

REVVA [2, 28] is a methodology for Verification, Validation & Accreditation (VV&A) of simulation models and stems from the defence industry. This methodology is summarised and adapted to manufacturing simulation in [29]. In this methodology, a third step is added to the verification and validation steps: Accreditation/acceptance. This is the decision of certifying that a model is correct and valid, or the decision that there is sufficient evidence about a model being correct and valid to assume that it is correct and valid. In many cases, acceptance will suffice, but in some cases a formal accreditation is required due to legal or contractual obligations. The REVVA methodology distinguishes a number of roles, such as contextual user, subject matter specialist and simulation provider.

3.6. Documentation and Presentation of Results

Results need to be documented and presented in an appropriate way. The need for documentation is obvious – to understand later why certain decisions were taken, without having to repeat the simulations. It also allows for comparison of the real (system) behaviour and the simulated behaviour, which gives an indication of how well the simulation predicted reality.

It depends on the target group how results are presented best. For instance, many manufacturing experts would be quite happy to see most results presented in the form of tables or graphs, which are easier to document and retrieve than animations.

Animations can be very useful to explain simulation results to stakeholders with no (or a limited) engineering background, or for training purposes. However, a danger with animations is that some people may draw their own conclusions from an animation. Furthermore, people sometimes get lost in discussing details of the animation instead of focusing on the major results.

3.7. Simulation as a Tool for Decision Making Under Uncertainties

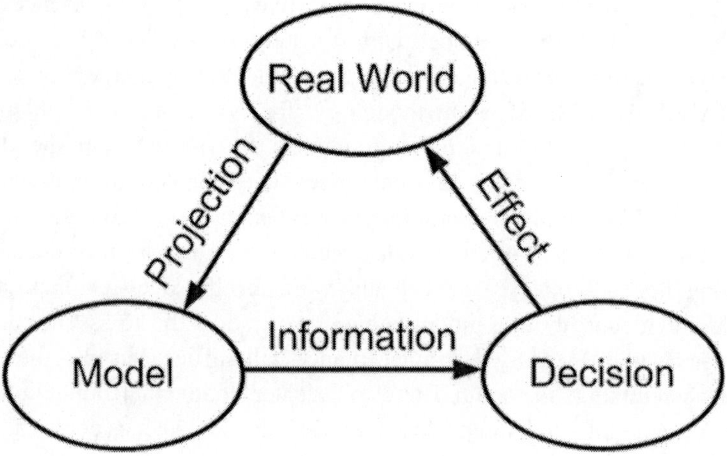

Figure 6. Interactions between Real World, Model, and Decision Arena. After [33].

In [30, 31, 32] it is explained why accurate knowledge about non-existing products is not possible; in essence, this is due to limitations of human perception. Likewise, behavioural indistinguishability between a model and its associated SoI cannot be demonstrated beyond all doubt. Hence, the simulated SoI may have other properties than the ones we envisage or expect. However, whilst simulation never can give a conclusive answer, it can still be used to reduce uncertainty. Strictly speaking, a simulation may not generate knowledge about the envisaged SoI, but it still provides information about the consequences that certain decisions would have if the model would be perfectly correct. For design choices or production solutions this often means that the decision made is sufficiently close to the optimum, under the conditions that the decision's robustness against variations is taken into account. Hence, the role of simulation is not to obtain knowledge about the envisaged SoI (as this is principally impossible), but to provide information regarding the (possible) consequences of certain design solutions. As such, it is a tool for informed decision making which helps the decision maker to analyze the potential effects of alternative decisions and to reduce the level of uncertainty in the decision making process. Figure 6 illustrates these relationships between Real World, Model, and Decision. [33] puts the use of models of an SoI in a concurrent engineering context which inevitably implies the use of models; for instance one cannot study the impact of the introduction of a new (to be developed) product variant in a production facility without using models of these. However, according to Fagerström, the nature of these models can vary; they can be mental models, mathematical models, 3D graphical models, and many more.

4. The Importance of the Human in Simulation Projects

Humans can play different roles in simulation projects. They can for instance be "subject matter specialist", "modelling & simulation expert" or "contextual user" [2, 28]. Humans in the first two roles play an important role in building simulation models and those in the third role typically are decision makers. In order to explain the importance of the first two, in Figure 7 below an expanded and refined version of Figure 1 is presented.

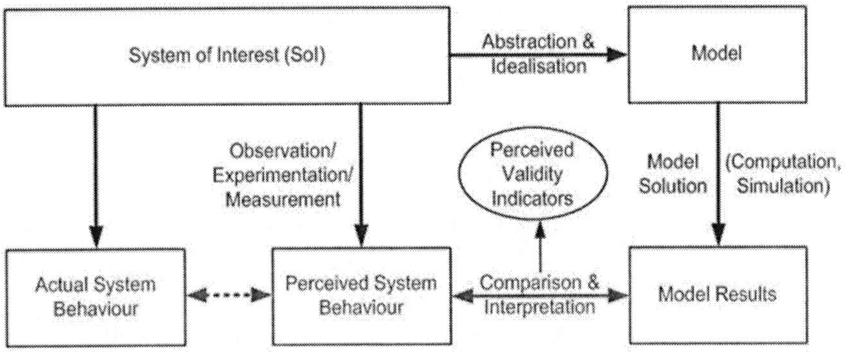

Figure 7. Refined view on Model and SoI.

A major difference between this representation and the REVVA representation from Figure 5 is that it is impossible to compare model results to actual system behaviour. This is due to principle limitations of human perception [32, 34]. As a result from this, the model results can only be compared to perceived system behaviour. This results in perceived validity indicators for the model. However, even this comparison is subjective in itself. A typical scenario would be that modeling and simulation (M&S) experts perceive a model as valid, whereas subject matter specialists perceive a model as flawed in some aspects. A possible remedy to this potential conflict of interests is to appoint independent/external VV&A executioners [2, 28]. For the contextual user, the most important validity indicator is whether the model results are an adequate basis for informed decision making. Since different types of decision require different information, to the contextual user (and thus, to other M&S stakeholders as well) validity of a simulation model and its results always have to be related to a specific purpose. Even when the contextual user is not an M&S expert, the user must have sufficient competence to order a simulation project or simulation model. This problem is addressed in for instance [35] in which a handbook for the use VM is described. The handbook consists of quick reference booklets, checklists, and more in-depth descriptions of the various simulation project phases. The handbook distinguishes between three main roles: production manager, project owner and simulation expert.

Apart from issues related to VV&A, M&S experts and subject matter specialists also have to address building a model of the right level of detail. Especially relatively inexperienced simulation model builders have a tendency to build models that are too large, too detailed and unnecessarily complex. This is partly due to the fact that this (building large and detailed models) is seen as a merit, and partly it is caused by a desire to satisfy subject matter specialists. Even if the subject matter specialists may not always say so explicitly, it is often thought that detailed models that accurately reflect reality will increase their tendency to accept the model as valid. However, detailed models take longer to build, they generally require more, and also more detailed data, and for some applications such as simulation-based optimization for on-line decision support, execution of detailed models may take too much time. They may also provide too many options to the contextual user and present simulation results with too much detail. As a result from this, the contextual user may find the simulation too complicated to run (too many options have to be set) or may find the results not clear enough for informed decision making. An example of a simulation tool for decision support that takes these issues into account is the FACTS Analyzer [36]. In FACTS, the underlying technology is fairly complex and incorporates algorithms from artificial intelligence as well as simulation models of various levels of detail, but the contextual user works with a relatively simple user interface.

5. Potential Pitfalls and Project Support

5.1. Potential Pitfalls

Unfortunately, many things can go wrong in a simulation project, and one could produce an extensive list of potential pitfalls without being anywhere near exhaustive. Some of the main pitfalls are listed below:

Model building and data-acquisition often take more time than planned, which results in too little time left for proper verification and validation. As an example, in a geometrical model one would not notice mixing up sine and cosine if one only tests with angles of 45°. Hence, a reduced verification and validation program can result in the wrong conclusions regarding the correctness of the model.

The project is ill-defined, or started too ambitious and simplifications/ assumptions are made along the way. This may result in the simulation becoming the goal, whereas common sense solutions or analytical models may work just as well (or even better), taking into account the simplifications made along the way.

Data is not scrutinized critically. As discussed above, acquisition of accurate data can be a big problem.

The model's use is extended to address questions for which it was never designed. The risk for this is especially high if simulation is carried out by a handful of enthusiasts in a company who seek management support for more widespread use of simulation. Once a simulation produces useful results, they are unlikely to turn down "can you simulate this as well" type of requests from management.

An old model is dusted off for later use without incorporating changes made to the real system. This could be due to poor documentation, but whatever the reason for this, almost certainly the old model will not produce reliable results.

The simulation is used to prove something that was already known. In this case, the simulation has no added value. Furthermore, there is a risk that one fiddles with the model until it gives the known "desired" result, or that one skips validation if it produces the "right" result at the first run.

People may draw their own conclusions from the animation. This is discussed above, and a simulation engineer should be very keen to explain limitations of the simulation/animation.

Simulation is no replacement for understanding or engineering knowledge – if you don't understand what you simulate, how can you understand the results? A typical pitfall here is to extrapolate results beyond the original scope of the model.

No, or poor sensitivity analysis. Poor analysis of the influence of different data sets on the results, or a tendency to carry out a lot of runs near what is thought to be the optimum, but which may be a local optimum. A related problem is poor knowledge of the length of the transient period of a model.

5.2. Project Support

The examples above show that many things can go wrong in simulation projects, even when experienced engineers and managers are involved. One of the problems identified is the lack of simulation project support, which can result in an unclear division of tasks/responsibilities and ad-hoc solutions to organizational issues. Although a number of good simulation handbooks exist, these handbooks are not suitable for daily support in simulation projects. In order to address this problem, the dAISy project was initiated. This project was industry-driven and focused on Discrete Event Simulation as a simulation tool commonly used in large corporate and SMEs. In the project, a common simulation methodology was developed and documented in a handbook [35]. The structure of the handbook is divided into a number of booklets, see Figure 8.

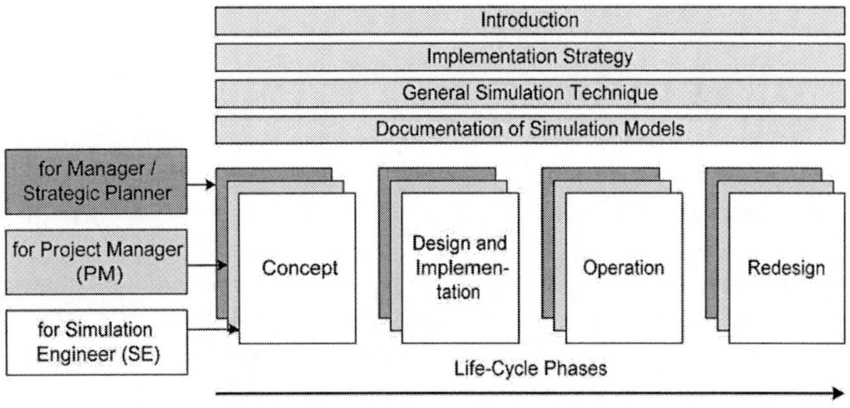

Figure 8. Structure of the handbook divided into booklets.

The booklets in the handbook are:

Introduction: The Introduction booklet describes the overall goal for the dAISy project and the handbook. It also describes the structure of the handbook, how to use it, definitions, etc.

Implementation strategy: Describes strategies, prerequisites, requirements and experience from implementation work (pitfalls, etc.).

General Simulation Technique: This booklet provides the reader with general knowledge about definitions, modeling, and the general sequence of activities in a simulation project such as data collection, verification, validation, etc. This booklet can be regarded as a textbook for simulation to provide the reader with some general knowledge about simulation. For a project manager it should be sufficient information to be able to purchase a simulation model.

Documentation of Simulation Models. The booklet not only addresses the documentation of simulation models but also full documentation of the entire simulation project.

Concept, Design and implementation, Operation, Redesign: life-cycle phases of the manufacturing system. The phases are adopted and composed from GERA. This booklet describes crucial activities in every life-cycle phase. The activities are decomposed into; inter alia, purpose, description, checklist and output.

Three different roles have been identified for successful integration and usage of discrete event simulation:

- Manager, strategic planner: responsible for integrating and developing new methodologies. For small and medium size companies this role is often dedicated to the vice president of the company.
- Project manager: responsible for ordering/buying a simulation. The project manager defines the task depending on the project's comprehensive goal. In general, this person is a well-qualified manufacturing engineer.
- Simulation engineer: responsible for the actual building of the model. The simulation engineer generally also conducts the experimentation on the model. This role is a simulation expert within the company or a simulation consultant.

The handbook is divided into guidelines for these different roles and different phases of the manufacturing system's life cycle. This division keeps the focus on the goal for the different roles also making it faster and easier for the user to find the guidelines for their specific role. The handbook presupposes that the simulation project is carried out as a sub- project within the framework of a larger industrial engineering project, which could have its own project manager. Each section consists of quick-lookup check lists and more extensive descriptions. The quick-lookup lists also provide more transparency about the other roles in the project.

6. Examples of Advanced Applications of Virtual Manufacturing

Advanced applications of virtual manufacturing include tools and techniques such as simulation-based online planning and scheduling, simulation-based remote monitoring & diagnostics, and simulation-based multi-objective optimisation. Examples of the latter are given below.

6.1. Simulation-Based Remote Monitoring and Diagnostics

The VIR-ENG project [37, 38] has highlighted the potential role of virtual engineering in machine system design. Whilst VIR-ENG itself focused on machine system development, it also forms the basis for the simulation based machine service support system mentioned below. The main objective of the project was to develop highly integrated design, simulation and distributed control environments for building agile modular manufacturing machine systems which offer the inherent capacity to allow rapid response to for instance product model changes. In the project, a component based paradigm was adopted for both hardware and software development. In essence, machine systems including their control system are developed in a virtual environment and subsequently implemented as a physical system.

In the MASSIVE project at the author's laboratory, a simulation supported machine service support system (MSSS, Figure 9) has been designed and implemented [39]. The system consists of on-site components, mainly for data acquisition (shown to the left) and components at the service provider site. The latter consist of components for data communication, storage and processing, components for analysis, and a user interface.

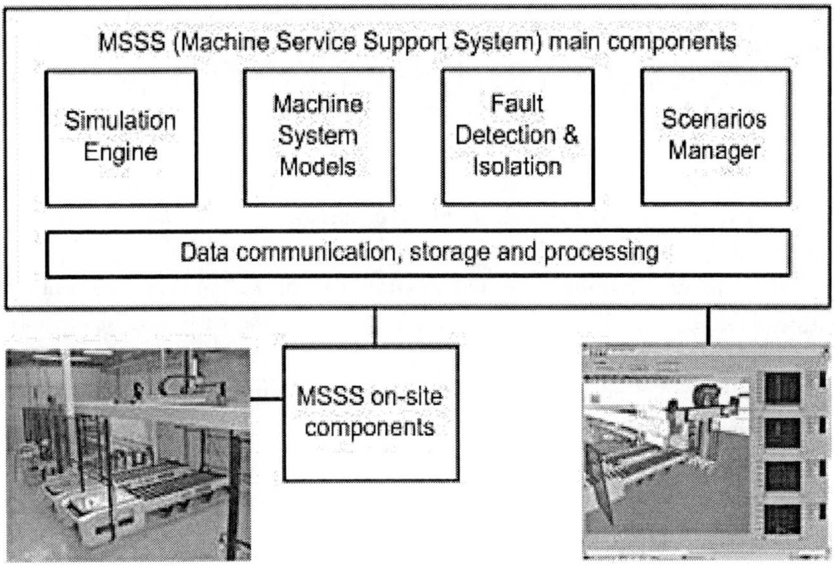

Figure 9. MSSS main component groups.

The approach taken in the project builds upon the tools and techniques from the VIR-ENG project and is an extended part of the machine system design and control environments from VIR-ENG. A key element in this approach is the seamless integration between simulation model (which is a hybrid DES/CAR model although CAR prevails) and physical equipment. This integration makes it possible for instance to:

- Study failure modes and their effects during machine system design through simulations in which certain disturbances/faults are emulated.
- Monitor the operation of the machinery system on-line, which facilitates both supervision by humans and data-acquisition.
- In the case of breakdowns, retrieve control code execution, sensor information from a temporal database and carry out a "replay" of the machinery system's recent history in the simulation model.
- Develop, test and upload temporary control code in the case of temporary reconfiguration due to machine service activities (this is a functionality inherited from VIR-ENG).

The kernel of MSSS is essentially a remote data acquisition and analysis system. An advanced data acquisition, pre-processing and management framework is the foundation for all other functions. The data acquisition system can be remotely configured so that specified parameters, machine process variables and signals can be acquired in prescribed time intervals and sampling rates. Configurations for routine periodic data logging can also be selected for day-to-day monitoring. Configuration of the data acquisition components is enabled through XML Web services using the user interface functions provided by the Scenario Manager.

MSSS offers continuous visual monitoring, but in the case of a machine failure (breakdown), MSSS users can also use the historical data saved in the database to carry out a "replay" to investigate the recent history of the machine system and current status using the corresponding simulation models. In these cases, animations are driven by the historical data acquired. Simultaneously, the reference process models are used to generate the nominal dynamic response of the system with the input data from the historical data. The output data generated by the simulator and from the collected historical data can be visualised and compared using various data analysis and residual analysis techniques. The data visualisation features enhance the 3D animation by presenting useful "non-animated" data like electric current and voltage produced both from the simulator and the collected data as an additional means for assisting any monitoring and diagnostic tasks. Fault alarms can be generated by the diagnostic agents of the system, for instance, if a residual signal is evaluated to exceed a certain threshold; but more advanced fault detection algorithms can relatively easily be incorporated into MSSS.

The remote monitoring function in combination with the replay function is very useful in a number of situations, in particular when the manufacturing system is installed at a remote location and the machine builder needs to know which type of service specialist needs to be flown in or which spare parts need to be sent to the site. When an error occurs occasionally, seemingly without a pattern, it can be difficult to decide what causes the error. With the use of MSSS, the pattern can be unravelled by running replays using media functions such as “slow motion” “rewind” and “fast forward” and using the scenario manager to select data signals of particular interest. In this way, “situation awareness” about the machine system's condition and the errors can be obtained. The user can select different views and signals to be shown, including overlay plots of signals. For cells consisting of multiple machines, a dummy device can be used if the internal clock of the simulation software does not have a sufficiently high resolution for time-stamping and synchronisation of signals [40].

6.2. Simulation-Based Optimisation

Another advanced use of simulation is to use it in simulation-based optimization (SBO). Simulation in itself is not an optimization tool, but it can be used to evaluate candidate solutions generated by a human or automated tool. Traditionally, candidate solutions are generated through Design of Experiments (DoE), but this requires a user that is both a simulation expert and a subject matter specialist. Furthermore, for complex real-life problems, DoE becomes a very elaborate and time consuming task. An alternative approach that is more user-friendly for the contextual user is to generate candidate solutions with the use of soft computing techniques such as Genetic Algorithms. These candidate solutions are then evaluated using simulation; the results obtained from this evaluation are subsequently used to generate new solutions (Figure 10). In this way, a set of candidate solutions can be presented to the decision maker.

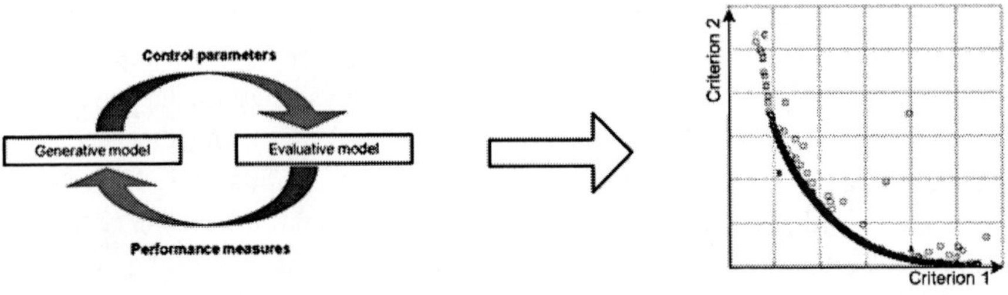

Figure 10. Principle of Simulation-Based Optimisation (SBO) with sample output (Pareto front).

In a project called SIMBOseer [41] this approach is demonstrated for the generation of machining sequences for deburring of engine blocks by two robots (Figure 11). The operations have to be divided over the two robots and their sequence planned under a number of constraints, such as the reach of each robot and blocking rules (robots cannot machine one surface at the same time) In that particular application, various scenarios were studied such as introducing engine block variants (requiring more, or fewer operations) in the production line. In such cases, the simulation-based optimisation platform was able to improve solutions generated by experienced process planners with 7% to 8%.

In most real-life problems, there are several types of objectives that have to be addressed simultaneously. This usually means that different objectives are given different weight factors when searching for an "optimal" overall solution. Moreover, many real-life problems are not static but dynamic, meaning that the relative importance of objectives such as environmental impact, production costs, and delivery time may vary. Hence it is necessary to generate a set of near-optimal solutions that can be selected depending on the situation at hand. This, multi-objective optimization, can also be achieved with the use of simulation tools, similar to SBO described above. For large and complex problems, this may require computing power that is beyond the resources of the contextual user's organization. However, this can be solved by running the actual simulation and optimization on a server cluster, with a client computer application at the contextual user end [36]. Due to uncertainties in the input variables, for instance energy prices, a post-optimisation analysis is usually carried out in order to present only candidate solutions that are sufficiently robust to the decision maker.

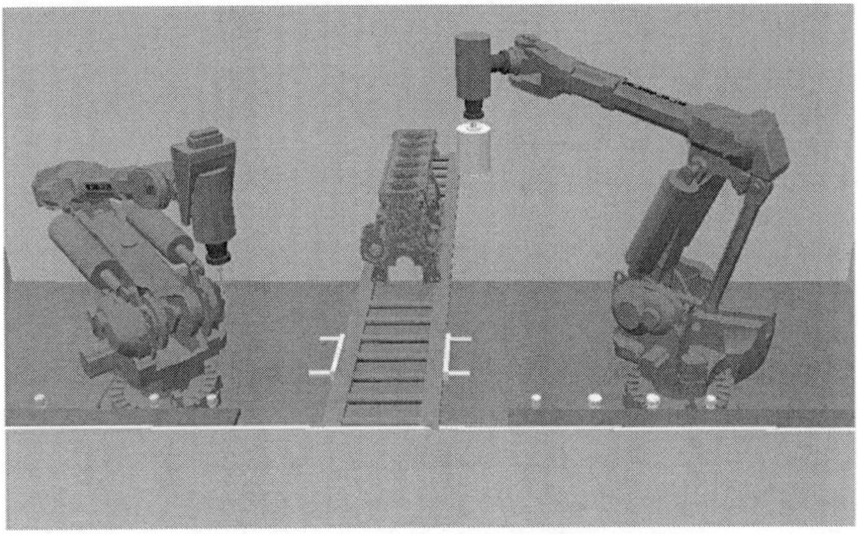

Figure 11. Robot cell studied and optimised in the SIMBOseer project.

7. Information Fusion and Models for Information Fusion

Many simulation projects aim at providing decision support for human decision makers. In a simulation project, often data from a number of different sources (for instance different in terms of time, modality and level of detail) is used and aggregated to understand for instance the behaviour of a production line and to provide the decision maker with a "what-if" analysis of alternative candidate decisions. This motivates extending our view by comparing simulation with information fusion.

7.1. Information Fusion

Information Fusion (IF) encompasses the theory, techniques, and tools conceived and employed for exploiting the synergy in the information acquired from multiple sources (sensors, databases, information gathered by human, etc.) such that the resulting decision or action is in some sense better (qualitatively or quantitatively, in terms of accuracy, robustness, etc.) than would be possible if these sources were used individually without such synergy exploitation [42]. An example of IF in manufacturing is the fusion of information from multiple sensors [43]. Examples of IF in the context of this chapter can be the fusion of information from the past operation of a manufacturing system (e.g., stored in databases), from the present (e.g., sensor signals, machine status), and from the future (in particular, predictions obtained through simulations). High level IF often has the purpose to provide decision support to a human decision maker (or group of decision makers). In this respect, it has similarities with modelling and simulation for decision support, as will be explored further on in this chapter. The two most prominent models for IF are the OODA Loop and the JDL model. Both models have their origin in defence-related work.

7.2. OODA Loop

The OODA Loop, also known as The Boyd Cycle [44, 45, 46], stands for Observe, Orient, Decide and Act. The model describes high level human-in-the-loop IF. It is, to some extent, comparable with Deming's PDCA circle; both can be seen as models for change management [47]. Despite the model's background from defence research, it can be used to describe many other decision processes. For an army officer, enemy troop movements possibly imply a threat. For a production planner who receives a phone call from a salesperson out in the field that a big order can be acquired if the company would be able to start delivering the next week, there is clearly an opportunity. But basically, these two problems are of the same nature: A new situation is developing, and within a limited timeframe, a decision, based on the information available just then, has to be made on how to address this new situation with limited resources.

7.3. JDL Model

A model initially developed for data fusion (an important sub-area of information fusion) is the so-called JDL model. In 1991, the Joint Directors of Laboratories in the U.S., with input from the information fusion research community's leaders, developed a data fusion paradigm. This paradigm, shown in Figure 12, aimed at providing a framework for communication and coordination amongst the many diverse fusion workers. The initial names of the levels are based on the initial military defence context, however a more general terminology is proposed in [48]:

Level One Fusion Processing – Object Refinement. Level one processing combines parametric data from multiple sources to determine the state and other attributes or identity of low level entities.

Level Two Fusion Processing – Situation Refinement. Level two processing develops a description or interpretation of the current relationships among objects and events in the context of the operational environment. The result of this processing is a determination or refinement of the operational situations.

Level Three Fusion Processing – Strategic Refinement. Level three processing develops an extra-organizational oriented perspective of the data to estimate extra-organizational capabilities, identify opportunities, estimate extra-organizational intent, and determine levels of risk.

Level Four Fusion Processing – Process Refinement. Level four processing monitors and evaluates the ongoing fusion process to refine the process itself, and guides the acquisition of data to achieve optimal results. This includes interactions among the data function levels and with external systems or the operator to accomplish their purpose.

Level 2 is increasingly also referred to as "situation awareness" and Level 3 as "impact analysis". For instance, presentation of results as in Figure 10 could be part of "impact analysis". Usually, a Level 0 Processing is also distinguished [49]. This Level 0 Processing deals (in military terms) with sub-object data. It incorporates for instance data analysis, resolving data conflicts, and conversion of data (for instance, to a common format or to common time & location frames). Llinas elected not to incorporate a Level 5 Processing (User Refinement) in his 2004 paper [49], as this was not, at least at that time, an established extension within the Information Fusion research community. However, within the context of this chapter, Level 5 is highly relevant and will be discussed briefly below.

Level 5 Processing as proposed by Blasch and Plano [50] in their JDL-U (with "U" standing for "user") model includes determination of who has access to information or queries information, and presentation of information to support cognitive decision making and subsequent actions. Important factors in Level 5 processing are workload, attention (including guiding and controlling attention), and trust in automated systems. Humans typically generate a mental model that describes purpose and explanations of system functions from observed states, and predicts future states. Recurrent situational scripts and scenarios are experiences that will contribute to the development of the human from novice to expert.

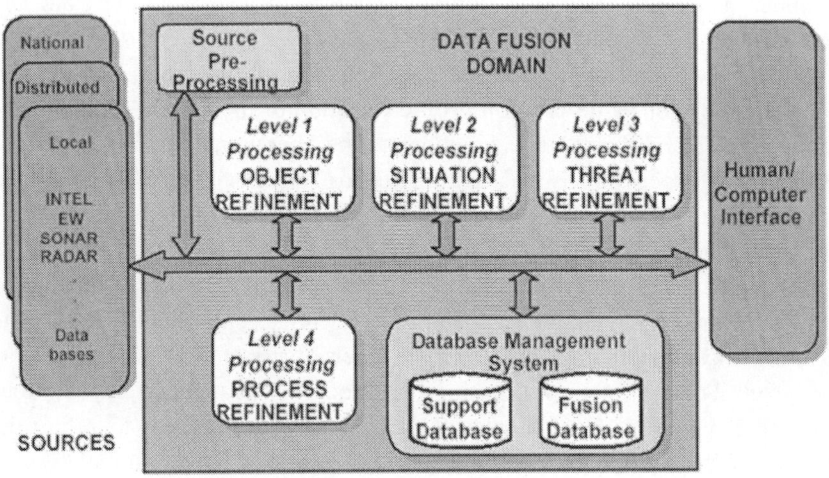

Figure 12. The original JDL Model.

8. THE JDL-U MODEL AS A REFERENCE MODEL FOR SIMULATION PROJECTS

Below, it is discussed how the JDL-U model can serve as a reference model for simulation projects. In this chapter, the focus is on simulation as a process, meaning that the focus is on data, models, and project phases. It is also possible to use the JDL-U model to describe how the contextual user gains insight in the behaviour of the SoI during a simulation project, but that comparison is rather trivial.

8.1. Comparison between JDL-U Levels and Simulation Project Activities/Tasks

With the purpose of high level IF and M&S often being similar, namely providing support for informed decision making, it would appear to be obvious to compare IF and M&S. In essence, in both cases, there is a human in the loop decision process. However, even on the lower levels, the similarities are striking. Figure 13 shows how various activities in a simulation project relate to the six (0-5) levels of the JDL-U model. Since the six levels of IF have already been discussed before, below we name the level and then describe the corresponding level/activity in an M&S project.

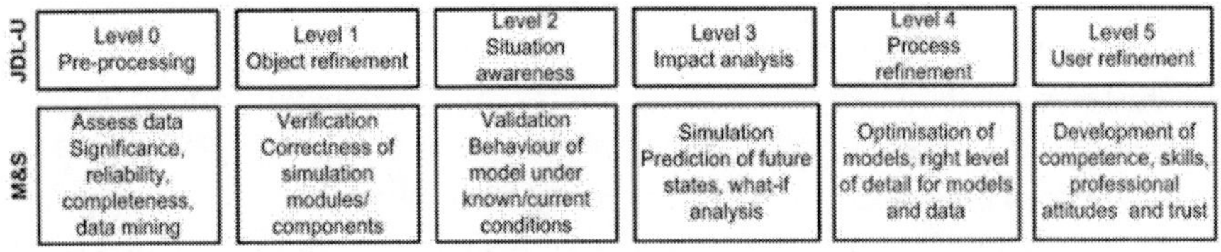

Figure 13. Comparison of JDL-U levels and Modelling &Simulation phases

Level 0 – Source Pre-Processing

The corresponding activity in an M&S project is very similar to that in an IF process, such as gathering and analyzing data, resolving conflicts, and removing multiple entries of the same data. Low level data mining can also be executed here to unravel patterns i.e. low level relationships between data. Data of less significance for the task at hand can sometimes be discarded.

Level 1 – Object Refinement

The corresponding activity in M&S is model verification, i.e. an analysis of whether the building blocks of a simulation model are correctly implemented. This can be compared to correct identification of objects (in the original military IF terminology).

Level 2 – Situation Awareness

The corresponding activity in M&S is model validation, i.e. an analysis of whether the model as a whole behaves in a way that is trustworthy. In essence, this means that the behaviour of the model is compared to the behaviour of the real system under controlled conditions, compared to theoretical behaviour (e.g., using trends and lower/upper bound analysis), or compared to results from a previously validated model (e.g., a model with a higher level of detail).

Level 3 – Impact Analysis

At this level, predictions about future states and their impact are made. The similarities between IF and M&S are most striking for discrete event simulation projects, in which different production layouts, different planning solutions, or the effects of the introduction of new products or product variants can be studied. However, more generally speaking, this level deals with prediction of future states and thus also accommodates for instance analysis of the impact of design solution on product properties/features. An example of this is the previously mentioned SIMBOseer project.

Level 4 – Process Refinement

Just like for IF, the first four (0-3) levels typically deal with single projects or scenarios, whereas Levels 4 and 5 mainly pertain to improvements made from one project/scenario to another. For M&S projects, this means building models of the right level of detail, increased insight into which data are crucial and which are less relevant, speeding up model building through modularization, and so on. Process refinement in M&S is thus fed by comparing simulation results with actual outcomes of implemented solutions, case studies in which alternative models over the same system are built, and so on. This level mainly pertains to M&S experts and to a somewhat lesser extent to subject matter specialists. However, contextual user aspects such as defining the right problem and ordering the corresponding simulation study can also be seen as part of process refinement.

Level 5 – User Refinement

Whereas Level 4 mainly pertains to M&S experts and subject matter specialists, Level 5 mainly pertains to both these and to the contextual user. An example of user refinement is improved distribution of tasks and responsibilities (although this can also be done within a single simulation project). Another form of user refinement is the creation of trust in simulation projects amongst the contextual users. In order to achieve this, simulation projects must yield results that to some extent are predictable, in the sense that similar studies should yield similar results. Apart from this consistency, the results should also form an adequate and dependable basis for informed decision making.

8.2. Examples of Level 1 and Level 2 Entities

Examples of entities on Level 2 can be production cells or lines in the case of discrete event simulation. This is in principle the level that is subject of the simulation study. The corresponding entities on Level 1, i.e. one level lower, are for instance machines or stations. In the case of computer aided robotics (geometry simulation), machines or stations are Level 2 entities whereas devices are Level 1 entities. In analogy with terminology from ISO 15 228, these can be seen as Systems of Interest and System Elements of Interest respectively. In principle, this can be a recurrent decomposition as shown in the line-machine-device example above. A corresponding example from product development could be a MBS (multi-body simulation) model incorporating condensed FEM (finite element modelling) models of flexible components which in turn have elements as sub-system entities. The Level 2 entities are also the entities that are of most interest for Levels 3-5.

8.3. Comparison on Level 3 and Refinement on Levels 4 and 5

Just like "the proof of the pudding is in the eating", the proof of simulation results is in how well they compare to observed behaviour of implemented solutions. In cases where simplified models were used, their suitability to offer adequate decision support can be judged by a comparison with results obtained from more detailed models – provided that these more detailed models have been validated previously. A critical assessment of simulation results and re-checking these against the problem and goal descriptions can result in Level 4 Process Refinement, such as determination of required level of detail of models, data, required number of iterations, or length of the transient state in case of simulation that need a "warm up" time. In the longer run, these assessments result in the development of human skills, competences and professional attitudes across various stakeholders. Professional attitudes include objectivity during validation phases, and correct use of simulation studies. For instance, a simulation study should be used to provide decision support, and not for retrospectively motivating an already taken decision [25]. Figure 14 illustrates how this process of critical comparison on Level 3 results in Levels 4 and 5 process refinement. Whereas Level 4 refinement is, to some extent, included in the REVVA representation of Model and SoI (Figure 1), the more detailed view presented in Figure 7 more clearly highlights the importance of Level 4 and Level 5 refinement.

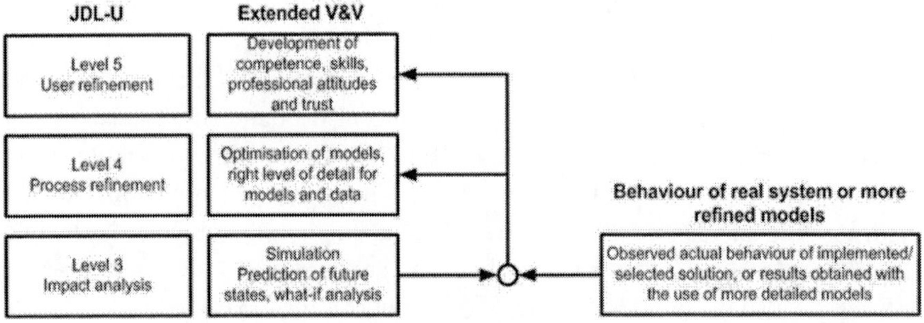

Figure 14. Comparison between simulated and observed results on Level 3 contributes to Level 4 and Level 5 refinement.

CONCLUSION

Virtual manufacturing is usually restricted to one or two domains in the PPR (product-process-resource) hub. A simulation project contains a number of phases and usually, a number of different stakeholders are involved, in different roles and often with different levels of experience with virtual manufacturing. This can make simulation project management a complex task, which this motivates the use of structured project handbooks such as dAISy.

Virtual manufacturing tools have also proven to be useful for a variety of advanced applications. One such application is remote monitoring and diagnostics, which allows service support to remote locations whilst only requiring modest bandwidth. Another application is simulation-based, multi-objective optimization for decision support.

A comparison between the activities in a simulation project and the levels in the information fusion reference model known as JDL-U show striking similarities between the two. From this, it can be concluded that the JDL-U model together with its associated science base is very apt to serve as a reference model for simulation projects.

BIOGRAPHICAL NOTES

Leo J. De Vin is a Professor holding the Chair of Automation Engineering at the University of Skövde, Sweden and an unsalaried Associate Professor in Computer Science at the University of Örebro, also in Sweden. He obtained an MSc in Mechanical Engineering in 1989 and a PhD in 1994 on the topic of

Computer Aided Process Planning from the University of Twente, The Netherlands. After working as a Post-Doc researcher at the University of Ulster in Jordanstown, Northern Ireland, UK, he joined the University of Skövde as a Visiting Lecturer in Integrated Product Development. He is a founder of the Virtual Systems Research Centre at the University of Skövde, who are Preferred Research Partner under the Volvo Group Academic Partner Program in the area of Virtual Manufacturing.

REFERENCES

[1] Von Karman, T., *Attributed quote*

[2] PROSPEC (2004), THALES JP11.20 Report JP1120-WE5200-D5201-PROSPEC-V1.3. Accessible at *http://www.vva.foi.se/revva_site/index.html*

[3] NSF (2006), Simulation-Based Engineering Science, *http://www.nsf.gov/pubs/reports/sbes_final_report.pdf*

[4] De Vin, L.J. and Sohlenius, G. (2006a), *The Role of Simulation in Innovative Industrial Processes,* IMC-23, Jordanstown UK, pp. 527-534

[5] Morris, A.J. and Vignjevic, R. (1997), Consistent finite element structural analysis and error control, *Comput. Methods Appl. Mech. Engrg* 140, 87-108

[6] Sellgren, U., (2002), Component Mode Synthesis - A method for efficient dynamic simulation of complex technical systems. Technical Report, KTH Stockholm

[7] Vreede, P., (1992), *A finite element method for simulation of 3-dimensional sheet metal forming.* PhD Thesis, University of Twente, The Netherlands

[8] Wiklund, D. (2006), Tribology of Stamping - The Influence of Designed Steel Sheet Surface Topography on Friction, PHD Thesis, Chalmers University of Technology

[9] De Vin, L.J. (2005), *Air Bending of Sheet Metal, FAIM 2005*, Bilbao, Spain, pp. 465-473

[10] Duflou, J.R., Váncza, J. and Aerens, R. (2005), Computer aided process planning for sheet metal bending: A state of the art, *Computers in Industry* 56, 747-771

[11] DELEM, (2006), *Private communication*

[12] Atzema, E., (1994), *Formability of Sheet Metal and Sandwich Laminates.* PhD Thesis, University of Twente, The Netherlands

[13] Klingenberg, W., Singh, U.P. and Urquhart, W. (1994), A Finite Element Aided Sensitivity Analysis of the Free Bending of a Drawing Quality Steel, *Proceedings of the 2nd International Conference on Sheet Metal, University of Ulster*, pp 41-48.

[14] De Vin, L.J., Streppel, A.H., Singh, U.P. and Kals, H.J.J. (1996), A Process Model for Air Bending, *Jrnl of Materials Processing Technology* 57(1-2), 48-54.

[15] Lutters, D., Streppel, A.H., Huétink, H. and Kals, H.J.J. (1995), A process simulation for air bending, in: Proceedings of the 3rd International Conference on Sheet Metal, Birmingham, pp. 145–154.

[16] Lutters, D. Streppel, A.H. and Kals H.J.J. (1997), Adaptive press brake control in air bending, in: Proceedings of the 5th International Conference on Sheet Metal, Belfast, pp. 471–480.

[17] De Vin, L.J. and Singh, U.P. (1998), Adaptive Control of Mechanical Processes: Brakeforming of Metal Sheet as an Example, *Mechatronics98,* Skövde, Sweden, pp 141-146.

[18] Klingenberg, W. and Singh U.P. (2006), Further observations and review of numerical simulations of sheet metal punching, *Int J Adv Manuf Technol* 30, 638–644

[19] Singh, U.P., Maiti, S.K., Date, P.P. and Narasimhan, K. (2000), Numerical Simulation of the Influence of Air Bending Tool Geometry on Product Quality, Proceedings SheMet 2000, Birmingham UK, pp. 477-488

[20] Singh, U.P., Urquhart, W. and Miller, P.P. (1992), Finite element simulation of cutting performance of CNC punches, Proceedings SheMet 1992, Birmingham UK, pp. 125-140

[21] Urenda Moris, M., De Vin, L.J. and Eriksson, P. (2004), Introducing discrete event simulation for decision support in the Swedish health care system, In Proceedings of the 2004 Western MultiConference: Health Sciences Simulation 2004 (Eds. Anderson J.G. and Katzper M.), San Diego *CA:SCS-Society for Modeling and Simulation International*, San Diego, CA., pp. 48-53

[22] Bäckstrand, G. Högberg, D., De Vin, L.J., Case, K. and Piamonte, P. (2006), *Ergonomics Analysis In A Virtual Environment*, Proceedings IMC23, Jordanstown, UK, pp. 543-550

[23] Tistrand, C. and Bodin, P. (2001), Development of simulation / offline programming for robot-assisted press brakes (in Swedish), BSc Project, University of Skövde

[24] Olsson, M. (2002), *Simulation and execution of autonomous robot systems.* PhD Thesis, Lund University, Sweden

[25] De Vin, L.J., Oscarsson, J., Ng, A.H.C., Jägstam M. and Karlsson, T. (2004*), Manufacturing Simulation: Good Practice, Pitfalls, and Advanced Applications, IMC-21*, Limerick Ireland, pp156-163

[26] Karlsson, J. and Samuelsson, F. (2001), *Simuleringsteknik i industriella sammanhang (Simulation techniques in industrial environment),* BSc Thesis, University of Skövde, (in Swedish)

[27] Banks, J., Carson, J.S. and Nelson, B.L. (1996*), Discrete-event system simulation,* Prentice-Hall, Inc., Upper Saddle River, New Jersey 07458, 2nd edition

[28] Brade, D. (2004), A Generalized Process for the Verification and Validation of Models and Simulation Results. Dissertation, Fakultät für Informatik, Universität der Bundeswehr München, *http://137.193. 200.177/ediss/brade-dirk/meta.html*

[29] De Vin, L.J., Lagerström, H. and Brade, D. (2006b), Verification, Validation and Accreditation for Manufacturing Simulation, *FAIM 2006*, Limerick, Ireland, pp. 327-334

[30] De Vin, L.J. (2007), The Role of Simulation in Engineering Design and Production Development Processes, *FAIM 2007*, Philadelphia, pp 107-114

[31] Sohlenius, G., Fagerström J. and Kjellberg, A. (2002), *The Innovation Process and the Principle Importance of Axiomatis Design*, ICAD2002, Cambridge, MA, Paper 018

[32] Jonson, E. (1953), What is truth? (in Swedish, original title: Vad är sanning), Ehlins handböcker

[33] Fagerström, J. and Moestam Ählström, L. (2001), Demands on Methods for Developing Work Focused on Concurrent Engineering, ICPR-16, Prague

[34] Chalmers, A.F. (1978). *What is this thing called science*? Open University Press

[35] Jägstam, M., Oscarsson, J. and De Vin, L.J. (2004), Implementation in industry of a handbook for simulation methodology, 37th CIRP Seminar on Manufacturing Systems, Budapest.

[36] Ng, A.H.C., Svensson, J. and Urenda Moris, M. (2008), Introducing Simulation-based Optimization for Production Systems Design to Industry: the FACTS Game, *FAIM 2008 Workshop Paper*, CD-ROM Proceedings pp 1359-1372

[37] Olofsgård, P., Ng, A.H.C., Moore, P.R., Pu, J., Wong, C.B. and De Vin, L.J. (2002), Distributed Virtual Manufacturing for Development of Modular Machine Systems, *Journal of Advanced Manufacturing Systems* 1(2), 141-158

[38] Adolfsson, J. Ng, A.H.C. and Moore, P.R. (2000), Modular Machine System Design Using Graphical Simulation, 33rd CIRP International *Seminar on Manufacturing Systems, Royal Institute of Technology (KTH)*, Stockholm, Sweden, pp. 335-340.

[39] Sundberg, M., Ng, A.H.C., Adolfsson, J. and De Vin, L.J. (2006), *Simulation Supported Service and Maintenance in Manufacturing, Proceedings IMC23*, Jordanstown, UK, pp. 559-566

[40] Pettersson, L., Adolfsson, J., Ng, A.H.C. and De Vin, L.J. (2007), Cell Monitoring and Diagnostics Using Computer Aided Robotics, *Proceedings of 40th CIRP International Seminar on Manufacturing Systems,* Liverpool UK,

[41] Holm, M., Doverborn, J., Ng, A.H.C. and De Vin, L.J. (2009), Optimisation of Operation Sequences in Flexible Manufacturing Cells using Virtual Manufacturing Tools, *FAIM 2009 Conference*, Middlesbrough UK, pp 1317-1324

[42] Dasarathy, B.V.(2001). Information Fusion – What, Where, Why, When, and How?. *Information Fusion*, 2, 75-76

[43] Dasarathy, B.V. (2003). Information Fusion as a Tool in Condition Monitoring. *Information Fusion,* 4, 71-73

[44] Boyd, J.R. (1986). Patterns of conflict, 1986, available from *http://www.d-n-i.net/boyd/pdf/poc.pdf*

[45] Boyd, J.R. (1995). The Essence of Winning or Losing, 1995, available from http://www.chetrichards.com/modern_business_strategy/boyd/essence/eowl_frameset.htm

[46] De Vin, L.J., Ng, A.H.C., Oscarsson, J. and Andler, S.F. (2006). Information Fusion for Simulation Based Decision Support in Manufacturing, *FAIM 2005 Special Issue of Robotics and Computer Integrated Manufacture*, 22, 429-436

[47] Lawson, H.W. and Senge, P. (2009*). A Journey Through the Systems Landscape,* unpublished draft, 2009

[48] McDaniel, D.M. (2001).An Information Fusion Framework for Data Integration. Thirteenth Annual Software Technology Conference "2001 Software Odyssey: Controlling Cost, Schedule, and Quality", Salt Lake

City, Utah. Also accessible at *http://www.silverbulletinc.com/downloads/McDaniel_r3.PDF, 2001*

[49] Llinas, L. Bowman, C., Rogova, G., Steinberg, A., Waltz, E. and White, F. (2004). *Revisiting the JDL Data Fusion Model II.* Fusion2004, Stockholm, Sweden, 1218-1230

In: Manufacturing Engineering
Editors: Anthony B. Savarese
ISBN: 978-1-61209-987-3

Chapter 2

ADAPTATION TO MARKET CHANGES IN PRACTICE: EVIDENCES FROM LATENT MANUFACTURING CLUSTERS IN TUSCANY

Francesco Rizzi and Francesco Testa
Scuola Superiore Sant'Anna, Istituto di Management, Pisa, Italy

INTRODUCTION

Today's economic downturn is providing decision makers with an understanding of the importance to dramatically speed up the capability to adapt local economies to market changes. To align production and consumption patterns to sustainable performances at local and global level is a priority for both public and private actors. According to that, since adaptation strategies encompass multiple interactions between individual and collective actions, the complexity of the challenge lies in how to coordinate and implement *micro-* and *macro-*level theories in practice.

How to benefit from synergies between regional transformations and corporate reorganization? How to make these synergies effective? For trying to answer these questions entails we developed an in-field investigation on how to invest in inter-public-private partnerships as well as in inter-organizational linkages.

Our time is characterized by accessibility and comprehensiveness of information and, in turn, by its effects on society. Social networks represent a new dimension where companies can rise consumers' consensus, gather rumors to orient strategies, add value to their products.

From an entrepreneurial point of view, social networks are increasingly expressing the capability to determine the value of a product as well as of the firm itself. Since material flows are no more the only determinant of market balance, virtual relations are shaking deterministic metrics for assessing production performances to their foundations. Accordingly, companies that aim at improving their productivity can no longer rely only on the traditional factors of production theorized in Fordism and post-Fordism theories.

Information activate knowledge, and knowledge lead to non-casual success. Thus, the way information circulates among market actors can contribute to determine durability and consistency of performances.

Engineer and product managers are usually called to support the sustainability of the company by optimizing processes within a specific techno-economic paradigm. To get the most from internal knowledge and from the exchanges of knowledge between firms represents a relevant challenge indeed. To remove the barriers surrounding a fair process of clustering interests, objectives, needs, resources and values can represent a potential way to implement effective adaptation strategies to market changes.

Grounding on a scientific literature review about the management of networks intended to improve the competitiveness in manufacturing, this chapter highlights some successes and failures in the adoption of the "Virtual Organization" as an organizational and production paradigm for coping with continuous changes, new environment and working conditions regulations, improved standards for quality, and fast technological mutation.

The discussion comprises two cases studies. These case studies, both carried out in Tuscany, are referred to locally well established but fragmented sectors that are asked to handle new market challenges: the food and energy sectors. Both these sectors, on the one hand, are expected to share competences for improving competitiveness and, on the other, are asked to share a way to overcome lack of effective approaches to interoperability. The first one provides the analysis with a meaningful context where barriers to change are shown on demand side, the second one with a context where barriers are shown on supply side. Finally, the evidences from case studies provide insight into the willingness to adopt new management models as well as into the strengths and weaknesses of public interventions to facilitate a systemic evolution.

In the first part of this chapter, challenges related to sustainability are presented as one of the major sources of pressures that encourage companies to enter external networks. After an overview of the key aspects that arise from the scientific debate, two real cases are discussed to highlight how difficult can be to align daily practices with theoretical predictions.

Stakeholders and Sustainable Production: The Role of Regulation

The production function is generally considered as being of interest for the parts (suppliers, producers and customers) that directly benefit from it. Despite that, since natural resources and common-pool resources, in general, can be impacted, the complexity of the net of agents that can determine its success tends to increase.

Industrial production is a major source of global pollution. At the same time it is widely recognized that regulation is required to reduce this pollution for the benefit of society as well as to preserve acceptability of well-managed activities.

Since there is considerable debate about the most effective approach to environmental regulation with respect to both environmental and competitive performance, a special attention has to be paid to the way productive organizations and stakeholders can express formal and bi-directional influences.

Regarding the different forms of environmental regulation, we can, to a first approximation, identify different categories of policy instruments depending on their ratio (e.g.: "Polluter Pays Principle" vs. market-oriented approach) and the degree to which they are compulsory:

- Direct regulation;
- Market based instruments.

Direct regulation (also called "command and control" regulation) includes standards as well as commands and prohibitions. One can distinguish between input regulation, process regulation, and output regulation.

Neoclassical theory suggests that command and control measures cause the highest cost of environmental policy instruments.

In the early 1970s, when environmental policies were still in their infancy, direct regulation was the most common approach of environmental policy, while economic instruments were used in only a few instances and were subject to much controversy. Since then a slow, but continuous evolution has taken place, with the role of economic instruments. First, the number of applications of economic instruments has increased as economic instruments are increasingly used in OECD countries. Second, the variety of instruments in use has also grown: while user charges and subsidies were already in use in the 1970s, different types of charges (e.g. emission charges) have become more common (OECD, 1999). Other types of economic instruments (e.g. deposit-refund systems, performance bonds, or liability payments) have also emerged. Another aspect of this evolution has been the growing role of environmental taxes, and the increasing number of applications of tradable permit schemes. The main instruments now in use for environmental protection are charges, environmentally related taxes, tradable permit systems, deposit refund systems, non-compliance fees, performance bonds, liability payments, and subsidies for environmental protection

Following a definition of the main economic instruments provided by the OECD in 1999, we can distinguish among:

- Emission charges: direct payments based on the measurement or estimation of the quantity and quality of a pollutant;
- User charges: payments for the cost of collective services. They are primarily used as a financing device by local authorities e.g. for the collection and treatment of solid waste and sewage water. In the case of natural resource management, user fees are payments for the use of a natural resource (e.g. park, fishing, or hunting facility);
- Product charges: applied to products that create pollution either through their manufacture, consumption, or disposal (e.g. fertilisers, pesticides, or batteries). Product charges are intended to modify the relative prices of the products and/or to finance collection and treatment systems;
- Taxes for natural resource management are unrequited payments for the use of natural resources;
- Marketable (tradable, transferable) permits, rights, or quotas (also referred to as "emissions trading") are based on the principle that any increase in emission or in the use of natural resources must be offset by a decrease of an equivalent, or sometimes greater, quantity. Two broad types of tradable permit systems are in operation: those based on

emission reduction credits[1] (ERCs), and those based on ex ante allocations ("cap-and-trade"[2]);

- Deposit-refund systems: payments made when purchasing a product (e.g. packaging). The payment (deposit) is fully or partially reimbursed when the product is returned to the dealer or a specialised treatment facility;
- Non-compliance fees: imposed under civil law on polluters who do not comply with environmental or natural resource management requirements and regulations. They can be proportional to selected variables such as damage due to non-compliance, profits linked to reduced (non-) compliance costs, etc;
- Performance bonds: used to guarantee compliance with environmental or natural resources requirements, polluters or users may be required to pay a deposit in the form of a "bond". The bond is refunded when compliance is achieved;
- Liability payments: payments made under civil law to compensate for the damage caused by a polluting activity. Such payments can be made to "victims" (e.g. in cases of chronic or accidental pollution) or to the government. They can operate in the context of specific liability rules and compensation schemes, or compensation funds financed by contributions from potential polluters (e.g. funds for oil spills);
- Subsidies: all forms of explicit financial assistance to polluters or users of natural resources, e.g. grants, soft loans, tax breaks, accelerated depreciation, etc. for environmental protection.

Market-based instruments include *economic instruments* such as duties, tradable emission permits and environmental liability (Hawkins, 2000), and *soft instruments* such as including voluntary industry agreements,

---

[1]This approach takes a "business as usual" scenario as the starting point, and compares this baseline with actual performance. If an emitter/user performs better than the anticipated baseline, a "credit" is earned. This credit can then either be used by the emitter/user himself, either at the current location or elsewhere, or sold to some other emitter whose emissions are higher than the accepted baseline (and presumably at a lower price than what it would cost the latter to abate on his own)

[2]The "cap-and-trade" approach sets an overall emission/use limit (i.e. the "cap") and requires all emitters to acquire a share in this total before they can emit. Emitters may be allocated their shares free-of-charge by a relevant environmental authority, or the shares may be auctioned. Regardless of how the initial allocation of shares is determined, their owners can then either utilize them as emission permits in current production, save them for future use (if "banking" is allowed), or trade them with other emitters.

communication and information measures as well as environmental certification schemes.

Soft instruments include *voluntary industry agreements*, *communication and information* measures as well as environmental management schemes. Strictly speaking, voluntary agreements, labelling and certification, and rewards are not economic instruments. However, they can play a role in achieving environment protection and its sustainable use by complementing and supporting the existing markets.

Environmental agreements aim to support organizations predisposed to negotiation and cooperation with institutions and other stakeholders, in order to promote implementation of environmental policies, that are integrative or incremental respect of actual local or organizations conditions.

Certification process scheme such as EMAS are based on the will of more active organization to demonstrate their own environmental commitment towards institutional interlocutors or commercial partners (Biondi et al. 2000). Information measures such as environmental label (European Ecolabel) aim to stimulate firms' innovation capabilities that are closer to final consumer.

All these regulatory tools tend to shift the attention of companies from a "win-loose" perspective (i.e. to reach the attended results with no matter for the external consequences) to a "win-win" one (i.e. to reach the results that are able to guarantee the replication of the performance). Consequently, companies are asked to consider relations with the surrounding environment and its actors as a potential source of competitive advantage.

The challenge for companies is to match internal and external expectations by assuring consistency and reliability of processes and products. Sometimes, this can be done through exerting efforts alone so as to achieve a complete differentiation with respect to competitors. Other times, this can be done through sharing knowledge, competences, resources or one or more factors of production with other companies so as to jointly strengthen the effectiveness of each one's actions and minimize the overall risks.

The paragraph that follows aims at depicting the importance of territorial proximity for succeeding while exerting productive and organizational synergies.

Organizations in Network: A Complex Dimension of Production Factors

The way companies compete in the market and for the market depends on the way other companies do it and on the capability of the hosting system to accept the overall perturbations (ecological economists refer to this capability as "carrying capacity"). Usually, the closer (in terms of siting and relations) companies are, the more they are able to adapt their strategies and production patterns to surrounding conditions. Accordingly, different forms of formal or informal, spontaneous or artificial, real or virtual aggregations (networks or clusters) of companies often catalyze a more efficient and sustainable adaptation by mean of synergies that conjugate expectations of the parts and of the whole.

Industrial districts and clusters are some of the more meaningful expressions of such aggregations.

The industrial district, by literature, is represented by a local system characterized by the presence of a main production activity performed by a group of small independent firms, highly specialized in different stages of the same production process. This peculiar entrepreneurial organizational model can develop synergies that result in a more efficient production than would occur within a single large plant. Marshall (1890) at the end of the XIX century had already highlighted the benefits coming from the agglomeration of economic activities in terms of availability of skilled labour and high level of specialization. Similarly, the Austrian economist Schumpeter in the first half of last century stated the existence of competitive advantages deriving from a business cluster. A significant contribution to the study of industrial districts and of internal relationship mechanisms able to generate competitive advantages for the cluster firms comes from G. Becattini who, in his article *"From industrial sectors to industrial districts"* introduced the concept of industrial district as a tool to support regional policies for territorial development.

Without any attempt to reorder the taxonomies in the field of industrial districts (which can be found in the literature according to the various configurations that such systems engage in), it is possible to find some definitions and acronyms that have slightly different interpretation and, as a result, partially conflicting definitions of local systems with a high concentration of businesses: from industrial districts to system-sectors, from *milieu* to TPS (territorial production systems), to the RESS (regional economic and social systems). What connects all the approaches underlying these concepts is the identification as the common element of the analysis a system of usually small and medium enterprises operating in a homogeneous sector (or in sectors known as "auxiliary") and located in a limited socio-territorial area in which they have deep-seated social and economic relationships. The role an industrial district can have in a competitive development of local production has stimulated in some national contexts the interest of *policy makers.* In Italy, Law n. 140 (1990), enacted to simplify and facilitate the set up of district areas, also fostered an institutional definition of the concept of Local Production System (LPS), which is an area characterized by:

Homogeneous production contexts;
High concentration of enterprises;
Specific internal organization.

Based on that concept, the industrial district can be considered a specific LPS featured by:

High concentration of industrial enterprises;
Highly specialized production of business systems.

When comparing the characteristics of industrial districts as previously mentioned with the concept of cluster at European level, three important common characteristics emerge. Firstly, clusters are seen as geographic concentrations of specialized firms, of highly skilled and capable workforce, of integrated production patterns and of supportive institutions that improve the flow and the spillover of knowledge. Secondly, the cluster is useful to reach the functional objective to provide a specific group of firms with a range of specialized and customized services. Finally, clusters are characterized by some social and organizational elements, called "institutional social cohesion tools", which link the different and interconnected actors, thus facilitating a closer cooperation and interaction between them.

The quality and quantity of the knowledge that circulate and the spillover among firms located in a cluster depend on the size of the cluster, on its degree of specialization and on how well the areas are equipped and focused on the production in the main industries that make up the cluster. Therefore, these three factors, size, specialization and focus can be chosen to assess whether the cluster has reached a “specialized critical mass” likely to spill-over and develop positive relationships. Statistical mapping of Clusters by the European

Cluster Observatory identifies over 2.000 regional clusters in Europe, among which clusters classified as “industrial districts” are 1.3803.

Even taking into account the different spatial dimensions of the European countries analysed, a greater presence of clusters emerges in Germany, Italy, Spain and France. Within the above mentioned countries, there is a predominance of the Construction and Food sectors with 15.5% and 10.7% respectively on the total number of districts. The Construction sector is more prevalent in Germany (27 out of 269 districts), in Italy (21 out of 158), and Spain (17 out of 104). In France, however, the greatest number of districts is in the Food sector, with 19 out of 103 districts.

In recent years, many initiatives have been implemented in Europe in order to create favourable conditions for the establishment of new clusters and strengthen existing ones. To date, more than 130 specific national measures in support of clusters were identified in 31 European countries and registered by the INNO-Policy Trend Chart4. Almost all European countries now have specific measures for clusters or programmes developed at national and/or regional level, suggesting that they are a key element of the national and regional strategies in support of innovation.

The local presence of natural resources can often foster the development of industrial clusters. In turn, the existence of industrial clusters is strictly dependent on their environmental sustainability. In fact, the size of clustering in a local context has critical relevance in the analysis of the environmental impact of industrial activities. When assessing the impacting factors related to a particular type of production, the characteristics of different local contexts in which that type of production produces its environmental effects have to be taken into account. Italy shows clearly how the environmental impact of some industrial sectors (e.g. textiles, tanning, ceramic) is localized around some areas where there is a high concentration of industries from those sectors. In these cases local dimension becomes a key determinant of the significance of environmental issues for the entire industry sector and, at the same time, a key variable in coordinating an effective response by the companies.

There is no doubt that in terms of impacts on the environment, companies that operate in an industrial district have many elements in common.

---

[3] Clusters that do not comply with the definition of Industrial Districts were not included in the total number of clusters considered by the European Cluster Observatory. Specifically, they are: Agricultural, Business services, Distribution, Education, Entertainment, Finance, Fishing, Hospitality, Sporting, Transportation.

[4] More detailed information in <http://www.proinno-europe.eu> and <http://cordis.europa.eu/ erawatch>

First of all settlement, production and sales activities of these enterprises influence the same local ecosystem, characterized by specific and defined environmental aspects. Moreover, companies operating in one district often face similar environmental problems, because they dump the emissions from their production processes into the same receptacle: waste water that drains into the same river (e.g. the Bisenzio river that runs through the entire Prato textile area, or the Sarno in the Salerno tomato district) or solid waste that goes into the same landfill.

On the other hand, the high specialization of production and the usually very small size of enterprises (with all the implications in terms of limited availability of human, technical and financial resources) allows us to think of the district as an industrial area sufficiently homogeneous also in terms of production methods, degree of technology and organizational and managerial choices. The same technological and organizational matrix of the businesses in the district may show in common environmental problems that are related, for example, to the inefficiency and ineffectiveness of facilities to reduce pollution, to technology obsolescence, to inadequate structures for environmental management, cultural lag and so on.

Even relations with suppliers of equipment and components, according to the logic of "vertically integrated industry" that characterizes many districts, are often played at local level, thus also affecting the availability and appropriateness of the most innovative and advanced technological solutions for pollution prevention (just think of the crucial role companies of the so-called mechano-ceramic have in the district of Sassuolo, being the almost only/exclusive repositories of technological know-how and, therefore, appointed to develop and propose new 'clean technology' to the ceramic businesses in the district).

A final aspect to highlight is the relationship with local stakeholders: for businesses in the district, interacting with the same community, the same institutions, the same local supervisory bodies means to deal with the same needs and requests concerning the quality of the environment. This is of fundamental importance if we consider that the significance of an environmental problem depends on the way in which it is perceived socially. The local dimension is a context where the relationship with company stakeholders is intensified, it becomes more straightforward (given the coexistence in the same area), more immediate (e.g. relationships with local institutions are more frequent than with national institutions), closer (just consider how much of the local population is employed by enterprises in the district). Besides, given the homogeneity of industrial activities, the physical proximity and frequent inability to attribute the environmental effects to any one production unit, enterprises in the district are considered by local partners almost as single entity.

The relational dynamics between companies and external stakeholders therefore become a crucial pressure factor to foster awareness on environmental issues within the district. By acting the same way and with the same incisiveness on a large number of similar businesses, it reinforces itself and strengthens its effects. For example, if the local population shows particular sensitivity to environmental issues, all enterprises in the district will undergo a high degree of examination from the public (and will therefore need to ensure continued compliance to regulations) and will be encouraged to use tools to enhance their environmental commitment to the local community.

Other important partners for companies in the district are local institutions. Sometimes companies interact with local authorities and supervisory bodies who are open to dialogue and willing to leave some room for negotiation, or with institutions that are particularly strict as regards law enforcement and extremely demanding on the compliance with obligations and deadlines. The different attitude of institutions can mitigate or amplify the context pressure acting in the same direction for all firms in the district. Firms can be challenged with requests from local authorities that may focus on some environmental aspects (making them more problematic) or that may promote the application of certain environmental policy tools (e.g. voluntary agreements at local level).

Local institutions may also prove to be particularly active in promoting common solutions (subsidiaries or consortium) to the most demanding and urgent environmental problems in the district, acting as a catalyst to encourage collaboration between businesses and promoting synergy in the commitment of human, technical and financial resources.

The local dimension represents an essential key in the understanding of environmental issues also because the same solution to environmental problems can be managed at district level. For example, the infrastructural equipment of a purification plant helps the industrial system in reducing the environmental impact. However, enterprises may find themselves having to directly invest in the installation of small treatment plants, which is known to result in a "scattered" distribution of facilities rather than in a systematic and consistent process.

Increasing awareness to environmental issues by the actors with whom the company interacts implies the need to meet certain “environmental questions”.

This is especially significant for SMEs operating within an industrial district. Efforts in the direction of environmental improvement by an individual company, in fact, are here associated with new knowledge and the onset of difficulties (the environment, as we have seen, is a challenge or new ”turbulence”) that once overcome constitute know-how that can be shared with other firms in the district. In this process of growth also appears the need for support from (and relationships with) external actors, a need common to most SMEs, which fosters the development of new ”answers” to the emerging needs. In a territorial dimension the resulting “networking” takes peculiar forms, leading to the development of somewhat common solutions (i.e. based on sharing tangible or intangible resources) that are tied to the specific local environment where businesses in the district interact.

Recent decades have shown the dynamic of those "common solutions" in industrial districts, connected to the different inputs and external forces that have enabled the development of strategies and tools to start up shared management processes that could involve the whole district.

Theoretical Perspectives: Trajectories from Lean Manufacture to Virtual Organizations

From an inter-temporal point of view, networks of companies are entities in continuous evolution.

While trying to act in the market in a systemic perspective, companies typically adapt their production and management patterns.

Usually, partnerships can assume the form of joint ventures, participation, clusters or virtual networks. The orientation towards one or the other of these forms seems to depend on the way the networks are formalized (i.e. by mean of equity or non-equity contracts) and on the prevalent shared goal (i.e. enforce each one's productions or coordinate them by mean of commercial agreements).

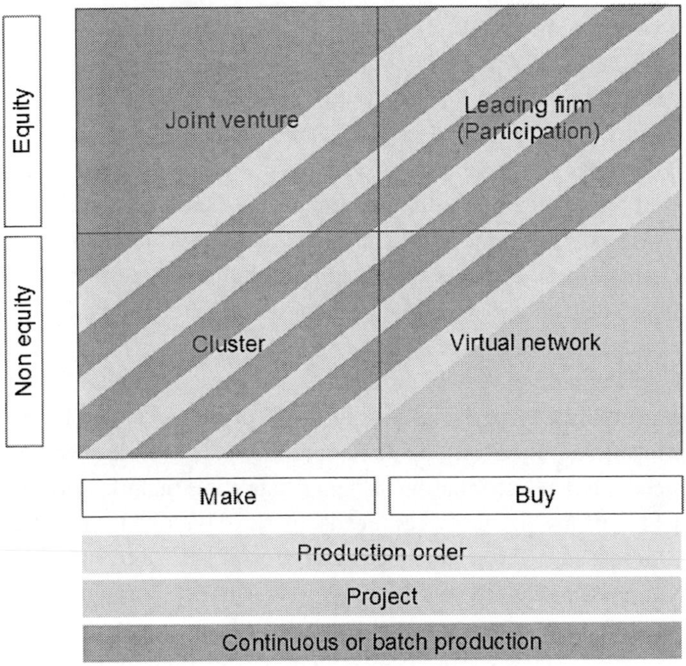

The production patterns of the firms involved in the network tends to favor the development of joint ventures for increasing volumes of production or for enter in the market with new products. On the other hand, single-project productions (customized productions) favor flexible virtual networks.

In this framework, the general tendency towards lean productions is often oriented to achieve the "six zero" goal: zero stock, zero defects, zero paper, zero waste of time, zero stops, zero conflicts. To achieve this goal, in turn, implies a tendency to substitute investments in plants with investments in knowledge and soft-wares. But the way these changes are implemented often depends on internal and external stimuli such as the availability of new ICT solutions, R&S achievements, partners' offers, customers' orientations, etc. Being these stimuli not completely predictable, companies can often seek greater advantages by investing in a systemic resilience rather than reacting to each specific event. To this end, to balance modular productions, open innovations, co-development partnerships and protection of critical knowledge can be assumed as the most relevant challenge for production managers.

In times in which businesses are restructuring and re-engineering themselves in response to the challenges and demands of 21st century, agility addresses new ways of running companies to meet challenges of demanding customers seeking high quality, low cost products, responsive to their specific and rapidly changing needs. Accordingly, agile manufacturing (AM) can be defined as the capability of surviving and prospering in a competitive environment of continuous and unpredictable change by reacting quickly and effectively to changing markets, driven by customer-designed products and services (Cho et al., 1996). Agile manufacturing, being a natural development from the original concept of "lean manufacturing", requires to meet the changing market requirements by enacting suitable alliances based on core-competencies, organizing to manage change and uncertainty and leveraging people and information.

In the AM paradigm, where multiple firms cooperate under flexible virtual enterprise structures, there exists a great need for a mechanism to manage and control information flow among collaborating partners. In response to this pressing need, an AM information system integrating manufacturing databases dispersed at various partner sites needs to be developed. Multipath agility provides increased access to alternative resources and information, and achieves improvements in productivity and quality through flexibility of access and utilization of resources, rather than through stepwise improvements in any one task (Gunasekaran, 1999). Here, companies must share the same goal as objective, create integration as the means, adopt IT as essential condition, negotiate key competence as key strategy (Jin-Hai et al., 2003). Furthermore, looking at the evolution from mass manufacturing to lean production, AM introduces the need to re-think the concept of competition and the importance of creating multiple-winners among manufacturers, suppliers and customers.

Despite the importance of cooperation between industries and academic organizations, many authors report an onward loss of importance of proximity and physical connections between cooperating actors that is caused by the ICT-improved efficiency and monitoring of this greater-than-the-sum-of-its-parts system (Scott Morton, 1990; Tapscott, 1996). Not surprisingly, from electronic trading on, the popularity of virtual structures is steadily increasing and legitimating the role of ICT as a driver for competitiveness (Davidow and Malone, 1992).

Impacts on productive units are relevant. In general terms, the increasing number of smaller firms globally, despite the breadth of advantages inherent to larger firms (Andersen, 2004), can be linked to the rapid technology transitions within global markets that facilitates the formation of SME networks introducing new organizational forms (Acs and Yeung, 1999) thanks to the substitution of physical burdens with "virtual linkages". In fact, even if - in situations that require long term and very large financial investments in infrastructure and R&D - multinational enterprises with large (tangible and intangible) capitalizations may remain the most efficient organisation form, as SMEs becomes more firmly networked and investors become comfortable with networked structures, combining assets of networks could provide equal sources of market capital access (Moore and Manring, 2009). By this end, virtuality in production patterns refers to a continuous process of negotiating among a group of firms so as to achieve the most effective way to deliver a product or a process.

Among the prominent definitions of "virtual organization" (VO) grounded on the analysis of such virtual linkages (Moshowitz, 1986) (Goldman et al., 1993) (Hardwick, 1996) (Upton et al., 1996) a major distinction arises between virtual collaborations within existing organizations (intra-organization form), also known as corporate network (Loebbecke et al., 1997), or between organizations (inter-organization form).

Inter-organizational virtual organizations, as temporary network-organizations, consist of independent enterprises that come together swiftly to exploit an apparent market opportunity. Here, virtuality relies on the ability of the organization to consistently obtain and coordinate critical competences through its design of value-adding business processes and governance mechanisms involving external and internal constituency to deliver differential, superior value in the marketplace (Venkatraman et al., 1994).

The degree to which virtuality is present on a variety of dimensions (competences, communication, etc.) varies with the contexts, and there is non single cut-off point at which a team becomes a virtual team.

In general terms, in a virtual organization, enterprises integrate vertically and externalize all but core activities and competencies. When new competencies are needed, they are added by acquisition or contract to a single organizational unit. Such unit replaces hierarchy-based controls in traditional organizations and can include sanctions for proscribed behaviour which are decided collectively and not on the basis of traditional partnership (bilateral) clauses. This negotiation phase is facilitated by information and communication technologies.

Reciprocal advantages are evident. As the network acquires the proprieties of a learning organization, by generating multiple feedback loops to the stakeholders, the emerging shared capabilities enable increased productivity. Due to the shared dimension of the goal, as a necessary condition for succeeding, individual SMEs must believe that the success of the whole network (mutual benefit) is paramount to individual SME success (Corral,

2002). Consequently, while assessing differential advantages, unlike in a company run on hierarchical principles, the variety of interests cannot be simply assessed on the basis of standard references such as return of investment but on the introduction of brand new competitive advantages, such as the market penetration through synchronized competency building (Manring et al., 2006).

To preserve that benefit, being a strategic partnership among stakeholders who come together to collaboratively address and resolve mutual concerns, the network has to provide the governance and structure to the dialogue and process, not the business. In fact, behaving more as a group of affiliates that a traditional supply chain network, the structure and processes of this superordinate entity directly impact stakeholders collaboration and consensus building efforts (Manring et al. 2006).

As well as trust is the basic ingredient of collaboration, trust is essential for the functioning and success of the virtual organization (Swagerman et al., 2000) albeit virtual organization represents a 'paradox' as it exhibits structural properties that work against the building of high trust relationships in which members' commitment to the virtual partnership is relational rather than purely transactional (Jones et al.,1998).

Among them, trust, both as defined as the expectation by one person, group or firm of ethical behaviour on the part of the other person, group or firm in a joint endeavour or economic exchange or as a psychological state comprising the intention to accept vulnerability based on positive expectations of the intentions of behaviour of another, is history-dependent and incrementally accumulated over time (Meyerson et al. 1996). Hereby, there is a trade off between the temporary and swiftly nature of virtual organizations and the time as prerequisite for trust building in business relationships which is solved by the possibility to build initial relationships (at design time) on high levels of trust. Such option is very important because trust contribute to risk mitigation and, fostering willingness to cooperate, reduces transaction costs which in turn increases the profitability and attractiveness of the virtual organization (Kasper-Fuehrer et al., 2001).

ICT, giving a contribution to bridge time and distance barriers, can facilitate a common business understanding (and transaction), which is a transient understanding between network partners. Therefore, for designing correctly the ICT support, the information exchanged can be broadly classified into transactional needed (e.g. to perform purchase, supply transactions, etc.) and managerial utilized (e.g. for decision making, control of business activities, monitoring goods, etc,) (Choe, 2008).

Theoretically, in a perfect VO there is no expenditure for maintenance of an organizational structure and thus, due to the lack of barriers to entry or exit, the only energy required is to find and accept connections to the network. On the other hand, there are times when a virtual network of expert units need more direction and guidance than a single unskilled team, if they have a tendency to take themselves in directions that they are either excited or comfortable with, and therefore not necessarily following the direction that the network needs them to take (Crossman et al., 2004). In this case, in the preparation phase, consultant serving as the net-broker or the facilitation provider, are often reported as useful to create a common bond and promote mutual trust. This caretaker function became a trust bridge (stakeholders trust the consultant before they really trust each other). In the same way, the place where initial meetings are hosted (e.g. municipal building) serves to promote trust. Anyway, membership should convey a sense of belonging to a community rather than a place.

To limit the costs of VO, it is important to reduce the negative effects of opportunistic behavior and the disadvantages of information islands. Accordingly, during maturity phase as well as during design phase, formal understandings about each member's role and evidences that the ethical standards are being adhered to together with actions that support group identification can determine critical behaviors such as willingness to cooperate with others. Over time a trusting atmosphere can encourage tolerance (e.g. in terms of less emphasis on formal coordination and compliance measures) and help keeping the alliance during difficult times. On the contrary, in worst cases, the lack of agreement to the equity of commitment given by the parties can cause the termination of the alliance despite the potential economic benefit (Crossman, 2004).

Planning the Revolution: Experiences in Food and Energy Sectors

In recent years, a large number of theoretical pathways to effective networking of manufacturers has been developed. Grounding on the discussion of in-field success-cases, best practices have been identified and conceptualized as well. Nonetheless, important contributions emerge also from failures. The cases that follow depict typical factors that have to be taken into consideration when top-down or bottom-up networking strategies are applied.

A Bottom-Up Approach: Evidences from Food Industry in Grosseto

Food industry has been traditionally considered as a low technology-intensive and technologically-mature sector (Grunert et al., 1997). Despite that, the increasing level of integration with upstream and downstream economic activities is introducing, directly, new challenges and, indirectly, new trajectories of innovation.

In Grosseto, the presence of an important food cluster reflects the agricultural vocation of the surrounding lands. In 2008 it was composed by a wide range of agricultural food chains (namely olive oil, wine, cheese, cereals, fruit and vegetables, chestnut) in which more than 200 companies (especially SMEs) operate. Beside these activities, as a consequence of the proximity to different Universities (Pisa, Florence, Siena), an ICT sector was present with more that 100 companies as well.

These two sectoral components, recognizing potential synergies in their reciprocal collaboration, actively cooperated in developing projects under regional funding calls (e.g. Docup ob.2 – 2000-2006, PRAI VINCI 2006/2007, etc.).

For ICT companies, to fasten relations with local agro-companies would mean to gain market shares from external competitors (mainly from multinational companies). For agro-companies, to faster relations with local ICT companies would mean to implement more taylor-made tools and to count on a more continuous and specialized assistance.

Hence, in past years, the collaborations between local agro- and ICT-companies, built on the clear vision that "to work together means to enforce local productivity, to develop unique inter-sectoral competence and, thus, to improve competitiveness", have been mainly and nominally focused on the mutual transfer of competences and, then, on the development of ad-hoc ICT solutions for food traceability.

In field of food traceability both agro- and ICT- territorial actors are involved in a cross-sectoral cooperation which is of great interest for local public administrations (see socio-economic impact) and is regulated by a complex negotiation process that is influenced - in turn - by a large variety of technical and managerial aspects. Under an entrepreneurial point of view, an efficient and effective system transmitting accurate, timely and complete information about products through the supply chain can significantly reduce operating costs and can increase productivity. At the same time, such a system contains many product quality elements. According to "supply-chain aimed traceability" (SCAT), this last point of view provides market leverage that companies can exploit to gain business advantages over competitors.

Beside competitive leverages, regulation compliance offers additional stimuli towards the adoption of traceability solutions. In food industry, the EU regulation no. 178, entered into force in 2002, makes traceability compulsory for all food and feed businesses. It requires that all food implement special traceability systems, in order to identify where their products have come from and where they are going and to rapidly provide this information to the competent authorities (one step-back one step-beyond approach). In addition to the general requirements, sector-specific legislation applies to certain categories of food products (fruit and vegetables, beef, fish, honey, olive oil) so that consumers can identify their origin and authenticity (the so called "consumer aimed traceability", CAT). The expected benefits for consumers are in terms of health, quality, safety and control.

The growing complexity and industrialization of food supply-chain has fostered also the development of wider approaches to traceability. Among them, voluntary schemes, as Italian standard ISO 22005, provide a higher degree of information associated with the individual product than the European mandatory traceability system and can combine CAT required by law with a SCAT approach useful for organizational and performance improvements.

In this framework, to reach the supply-demand equilibrium by counting on local firms would mean to improve local competitiveness. This is the reason why local public administrations are willing to play the role of facilitators in local business networking initiatives.

In Tuscany, local provincial administrations are highly-structured to promote this kind of projects. In 2007, de facto, also in Grosseto, the provincial administration took part to the activities by playing a role of institutional guarantor and coordinator. In particular, as a consequence of the bottom-up stimuli provided by ICT and agro entrepreneurs, the local provincial administration promoted the development of a Centre for the Food Quality as a physical space for catalysing activities aimed at, first, spreading the implementation of traceability procedures and, second, at establishing a quality label for local productions.

The idea was to develop a sort of virtual organisation among local ICT and agro players so as to facilitate, by means of a central coordination, negotiations and transactions of business opportunities. By means of focus groups participated by local administrations and entrepreneurs, the attention was focused on two critical thematics: the customization of ICT tools for cutting costs of traceability processes within agro businesses (e.g. taylor-made systems based on RFID technologies) and the integration of such tools with the implementation of active and intelligent packaging. Being the interest on these systems shared within clusters of homogeneous entrepreneurs (e.g. cheese makers, fish transformers, wine producers, etc.), the each one's investment on R&D was expected to benefit from economies of scale.

In 2008 a survey, conduced within the "Track" regional project, resulting in 100 agro manager and 38 ICT manager respondents, served the scope to better investigate the attitudes of local entrepreneurs and their willingness to play as leading actors in the development of the virtual organization (i.e. being interested in investing time and resources for catalysing the development of project-specific networks). As a result, local administration promoted public meeting so as to present the selected initiatives, facilitate and build trust in these cooperation processes. The first phase ended with the presentation of supply-demand postings and the arrangement of the first round of thematic meetings coordinated by leading partners.

Virtuality of the network was expected in terms of openness to new contributors, flexibility in negotiations and circulation of shared set of common design, engineering, and production efforts,

The following phase, the stage of maturity of the networking activities, was expected to serve as input for the fine-design of the Centre for the Food Quality.

In late 2009, an investigation on the selected actors (100 agro- and 38 ICT-firms) offered the opportunity for checking the progress of the activities of this virtual network.

Unfortunately, from this check emerged that only marginal clusters of actors operatively joined network's activities. After an in-deep round of interviews with involved actors emerged that, despite the general commitment to achieve the shared goal, the main killer factors the determined the failure of the initiative were:

The difficulties, for argo-firms, to run faster than the downstream players. After the first in-detail feasibility studies, agro-firms realized that they could expect benefits in terms of cutting costs for internal traceability (e.g. by substituting paper records with electronics devices), but could encounter troubles in relating with intermediate and mass retailers. In fact, in case of delivering information through intelligent packaging or high-capacity RFID codes, problems could arise in managing responsibilities among actors throughout the delivering chain and, than, in closing good deals with the actors that are asked to carry risks of generating non-conformities (e.g. mass retailers could prefer alternative products that are less expensive and more easy to manage). Additionally, without effective actions on customers, only marginal shares of the market are expected to understand quality information (e.g. about the correctness of conservation processes, of delivery times, etc.) and than, to prize them.

The difficulties, for ICT-SMEs, to enlarge the time-to-market of their services. Usually, local ICT companies are not structured to afford long-term investments and, thus, encounter greater difficulties in pushing products into the market than operating in a pull mode. As a consequence, as soon as agro-companies (ICT companies' customers) reduce their demand for highly customized products, ICT companies rapidly orient their R&D teams towards less challenging market niches. In turn, the more agro-companies have a reactive approach to sectoral regulatory and customer needs, the less ICT-companies are attracted in challenging co-development activities.

The limited availability of public resources to support the start-up and promotion of quality labels. Spreading awareness among customers on the intangible value – in terms of health, quality, safety and control – of purchasing products with a protected designation of origin requires large investments. Since local entrepreneurs have to contribute to this investments and, in the meantime, have to be attracted by the potential market, public investments play a critical role in building an adequate capital of trust and, thus, in orienting local firms' strategies.

To simplify, looking at the elements in common between these "quit conditions", the bottom-up intervention failed because of the lack of involvement of critical actors: citizens, intermediate and mass retailers. These critical actors are the ones that have to create (citizens) and allow (retailers) the demand for traceability and for intelligent and/or active packaging that is fundamental for feeding the network with motivation to invest in co-development activities.

A Top-Down Approach: Evidences from Energy Industry Niches in Tuscany

Since large private-public challenges, such the ones that entails environmental and health issues, require increasingly efficient multi-stakeholder dialogues, the Europe 2020 Strategy for a long-term sustainable future calls for action at all levels, including regions and in cities. Furthermore, policymakers agree upon the important role of regional and local authorities in driving green growth.

Along the chain from European, to national, to regional and, then, to local level, public authorities are committed to reduce CO2 emissions and fossil fuels consumption. In operative terms, a combination of command-and-control and voluntary tools is often considered to be effective for encouraging desired market behaviors. Among these tools, regional authorities often can count on the opportunity to found territorial projects that comply with environmental priorities. Here, multi-stakeholder trajectories grounded on the opportunity-driven unequalled added value provided by real-time field data managed through "machine intelligence" in a "smart service" perspective are often supported measures.

Hereby, in 2006 the Tuscany Regional Authority funded a project aimed at supporting the development of a local "hydrogen industrial chain". The project idea grounded on the presence of a significant intensity of H2-companies, mainly involved in production and supply of H2 streams and tools for metallurgic business, and of important research centres.

Hydrogen, even if it is not an energy source, but only an energy carrier, can play an important role in displacing emissions within highly polluted urban area (e.g. by replacing fossil fuels used for transportation) or in stocking extra-productions of energy so as to guarantee its availability during peaks without investing in new generation-plants (e.g. by developing the so called "smart grids"). Such uses are at the top of the priorities of future energy systems and, consequently, they are of great interest for both policy makers and manufacturers.

In 2006, regional H2 companies and research centres, albeit being both strongly oriented to compete in international markets, revealed modest interactions. In particular, upstream manufacturers (the ones that produce electrolyzers, industrial gases, high-pressure tanks and pipelines, fuel cells, etc.) were progressively shifting their market shares from local customers (e.g. metallurgic business operating in fashion and jewellery districts in Florence and Arezzo) to international customers. In the meantime, research centres were intensifying their cooperation with international players operating in field of automotive and ICT-UPS (Uninterrupted Power Supply) industries.

Despite the lack of coordination, it was internationally recognized that within Tuscany Region there is an wide range of competences and capabilities in H2-industry. Hereby, the goal that has been set by policy makers through a top-down approach was to favor the further development of such competences and, in particular, the sustain the improvement of the overall productive capacity in H2-industry.

Previous experiences in field of renewables, such as the introduction of feed-in tariffs for photovoltaic or green certificates for wind energy, were achieving significant results in terms of installed capacity, but without returns for local primary industries. In fact, while the sharp increase in demand was orienting investors towards the import of technologies from foreign manufacturers, local companies were gaining market shares only in terms of system integrators (i.e. downstream knowledge-intensive actors). Contrariwise, for H2-industry was perceived the opportunity to fully exploit the development of domestic technologies by grounding a progressive transition to an hydrogen economy on a joint push between policy makers, research institutes and – above all – local sectoral manufacturers. The presence of automotive and energy players, that are usually considered as latent catalysts of future H2-markets, was considered another important local asset.

Within this framework, different prototype developments were funded so as to create opportunities to link local companies and start co-development programs.

In late 2009, an investigation on 11 H2-companies, highly representative of the regional H2-industry, offered the opportunity for checking the ongoing returns in terms of increased willingness to cooperate among local manufacturers. The recorded outcomes are not positive as expected, but meaningful for further developments.

Local H2-SMEs, having the same background as technology providers for regional metallurgic clusters, are nowadays in a position of pure concurrence with each other. These manufacturers show a great attitude towards incremental innovation, but weak structure and resources to achieve radical innovations. As a consequence, their main factors of competitiveness consist of the capability to satisfy customers' needs in term of flexibility and customization. Unlike multinational manufacturers, that introduce into the market standardized products, local H2-SMEs operate in market niches where production is characterized by creativity and non-seriality. Being the creation of added value strictly linked to knowledge, they tend to protect it by co-developing products only with customers, not with potential competitors. Their customers, that are far from being mass retailers, in turn benefit from low-cost custom development services, timely and qualified lifelong assistance and low transaction costs. Hence, the strategic value of mutual trust in manufacturer-customer relations seems to counterbalance the windows of opportunities that are witnessed in higly-networked contexts.

Indirectly, research centres do not find in local manufacturers significant customers for their research activities and, thus, orient their efforts to meet the demand of big international players (e.g. multinational companies that contract R&D in electric vehicles or innovative UPS projects). Consequently, since a large number of local SMEs is not able to replace the demand of a single leading multinational, international tendencies drive the development of local capabilities more than local dynamics.

Local actors agree upon the usefulness of cooperation in prototype development for both co-financing investments at single firm's level and for sharing risks in testing new technology pathways, but are not willing to change their behaviors towards strategic knowledge management. Hereby, these top-down initiatives are considered useful to mitigate the impact of technology lock-ins and path dependencies, that means to improve the long-term resilience of local H2 economy, but also able guarantee only modest returns in the short term. From this point of view, the selection of the optimal innovation pathway appears strictly tied to a broad set of conditions that influence complex networking dynamics and that feature intra- or inter- sectoral, local or diffuse, transitional or permanent determinants.

As an example, local partnership can be developed under the stimuli of large firms (e.g. automotive manufacturers that are forerunner of hydromethane technologies or smart grid developers) only in case of non-serial production perspectives (i.e. the perspectives that do not lead to technology developments that are potentially less subject to imitation and replication by powerful multinational) due to the resistance of H2-manufacturers to accept the risks of competition for higher market shares. This means that, in such contexts, networking strategies can easily encounter bottlenecks on the supply side.

To simplify, when operating from a single order-out basis, open innovation, clustering or virtual networking are still possible, but negotiations over costs and benefits of sharing competencies can be perceived as a particularly critical issue (e.g. to fairly allocate the added value of creative developments among firms with similar core competences is perceived as difficult). Contrary to previous case study, to switch from a firm's push to a market pull environment would mean to increase transaction costs. To pursue network benefits, once more, strategies must be tailored to a critical mass of local manufacturers.

CONCLUSIONS

Nowadays, to compete in global market meeting the requirements of evolving (quality and environmental) regulation would imply to improve manufacturers' competitiveness by embracing networking strategies. Cluster approaches, open innovation and virtual organizations are some of the options that seem to fit better, on the one hand, with the culture of lean production and agile management and, on the other, with the increasing call for sustainable development.

Literature offers valuable insight for understanding the way firms can benefit from collaborative platforms. However, unfortunately, synergies and cooperation between manufacturers can not be easily planned.

As emerges from the investigations carried out in Tuscany on food and energy players, despite the support provided by local administrations, the willingness to adopt innovative management schemes can be wiped out by well-established competition practices. Such practices are rooted in equilibria that may be directly or indirectly controlled by external actors, as shown respectively in the first case study (where downstream players provide the market pull and, thus, determine the success of local initiatives) and in the second one (where large multinational confine local players to struggle for market niches).

In both cases, the lack of cooperation between companies entails that competences are distinctive and developed within each firm. In turn, due to the lack of a shared demand, no training programs are offered by local universities and training schools. As a consequence, the territorial intensity of distinctive skills is associated with the sense of belonging to single firms, not to a territorial system of firms. This, in the long run, does not turn to advantage of networking attitudes.

In conclusion, the conditions that led to the large presence (in numbers) of sectoral firms do not necessarily lead to further aggregation. Similarly, aggregation can not be simply determined neither by structural isomorphism nor by the sharing of market targets, being these conditions sometimes not sufficient (i.e. for food companies), and sometimes to-be-necessarily-avoided (i.e. for H2-companies).

In spite of that, the territorial intensity of sectoral firms often represents a determinant of territorial marketing and, in turn, the resulting brand of the territory (e.g. "a land of talented manufacturers") represents a shared trademark for sectoral firms. Being frequently capable of adding value to single firm's performances, such virtual cycle can be considered an intangible asset that deserve to be managed like short term hire. Hereby, further research will determine whether this could be considered the minimum shared target that local manufacturers could agree upon for jointly achieving their objectives in terms of growth and progress.

REFERENCES

Acs Z., Yeung B., (1999) *"Small and medium-sized enterprises in the global economy"*. Ann Arbor: University of Michigan Press

Andersen, E., (2004) "Population thinking, Price's equation and the analysis of economic evolution". *Evolutionary and Institutional Economics Review;* 1(1):127–48.

Biondi, V., Frey, M., Iraldo, F. (2000), "Environmental Management Systems and SMEs", *Greener Management International*, n.29, pp.55-69.

Blasch, E.P. and Plano, S. (2002). JDL Level 5 Fusion Model "User Refinement" Issues and Applications in Group Tracking, *SPIE Vol 4729.* Aerosense, 270-279

Cho, H., Jung, M., Kim, M. "Enabling technologies of agile manufacturing and its related activities in Korea", *Computers and Industrial Engineering* 30 (3) (1996) 323-334.

Choe, J.M., (2008*), Inter-organizational relationships and the flow of information through value chains, Information & management*, 45, 444-450

Corral CM., (2002) "Environmental policy and technological innovation: why do firms adopt or reject new technologies? *New Horizons in the Economics of Innovation",* Cheltenham, UK; Northampton, MA: Edward Elgar.

Crossman, A., Lee-Kelley, L., (2004) "Trust, commitment and team working: the paradox of virtual organisation*", Global networks*, 4, 375-390.

Davidow, W.H., Malone, M.S., (1992). *"The Virtual Corporation".* Harper Collins, NY.

Goldman SL (1993), Nagel RN. "Management, technology and agility: the emergence of a new era in manufacturing". *International Journal of Technology Management,* 8, 18-38.

Gunasekaran, A. "Agile manufacturing: A framework for research and development*" International journal of Production Economics*, 62, 87-105

Grunert, K. G., Harmsen, H., Meulenberg, M., Kuiper, E., Ottowitz, T., Declerck, F., Traill, B., & Goransson, G. (1997). "A framework for analysing innovation in the food sector". In B. Traill, & K. G. Grunert (Eds.*). "Product and process innovation in the food sector"*. London: Blackie Academic.

Hardwick M, Spooner DL, Rando T, Morris KC. (1996) "Sharing manufacturing information in virtual enterprises". Communications of the ACM, 39, 46-54

Hawkins, R., (2000) ,"The use of economic instruments and green taxes to complement an environmental regulatory regime", *Water, Air, & Soil Pollution* – Springer

Jin-Hai, L., Anderson, A., Harrison, R., (2003) "The evolution of agile manufacturing", *Business Process Management Journal*, Vol. 9, No. 2, pp. 170-189

Jones, T. M. and N. E. Bowie (1998) '"Moral hazards on the road to the "virtual" corporation*", Business Ethics Quarterly*, 8, 273–92

Kasper-Fuehrer E.C., Ashkanasy N.M. (2001) "Communicating trustworthiness and building trust in interorganizational virtual organization". *Journal of management.* 27, 235-254

Loebbecke, C., Jelassi, T., (1997), "Concepts and technologies for virtual organizing: the Gerling Journey*", European Management Journal*, Vol. 15, N.2, 138-146

Meyerson, D., Weick, K. E., & Kramer, R. M. (1996). "Swift trust and temporary groups". In R. M. Kramer (Ed.), *"Trust in organizations: frontiers of theory and research"* (pp. 166–196). Thousand Oaks, CA: Sage.

Moore, S.B., Manring, S.L., (2009) "Strategy development in small and medium sized enterprises for sustainability and increased value creation", *Journal of cleaner production*, 17, 276-282

Manring, S. L., Moore, S. B., (2006), "Creating and managing a virtual inter-organizational learning network for greener production: a conceptual model and case study*", Journal of cleaner production*, 14, 891-899

Moshowitz A. (1986) "Social dimensions of office automation". *Advances in Computers,* 25, 335-404

Scott Morton, M.S., (1990). *"The Corporation of the 1990s" Information Technology and Organizational Transformation*. Oxford University Press, NY.

Swagerman, D. M., Dogger, N., & Maatman, S. (2000). "Electronic markets from a semiotic perspective". Electronic Journal of Organizational Virtualness, 2 (2) 22–42. *http://www.virtual-organization.net.*

Tapscott, D. (1996). *"The Digital Economy: Promise and Peril in the Age of Networked Intelligence"*. McGraw-Hill, New York.

Upton DM, McAfee A. (1996) "The real virtual factory". *The Harvard Business Review,* 74, 123-133.

Venkatraman N, Henderson C. (1996) *"The architecture of virtual organizing: leveraging three independent vectors"*. Boston University: Discussion paper, Systems Research Center, School of Management.

In: Manufacturing Engineering
Editors: Anthony B. Savarese

ISBN: 978-1-61209-987-3

Chapter 3

CONVEYOR-LIKE NETWORK AND BALANCING

Masayuki Matsui
The University of Electro-Communications, Tokyo, Japan

ABSTRACT

This chapter focuses on two special queueing networks composed of a decision maker (coordinator) and K CSPSs (Conveyor-Serviced Production Stations), originaled in Matsui, 1982. One is a series' array of CSPSs (Model I), and the other is an ordered-entry array of CSPSs (Model II). The chapter first presents a station-centered approach to the class of CSPS Network, and prepares a few queueing formulas and general relation concerning average criteria. Next, a 2-level mathematical formulation of max-max and max-min types is explained and presented for determining both the cycle time and time-range (control variables) of optimizing the production rate of Models I and II ,respectively. Through the chapter, a mathematical theory (or method) concerning CSPS Network is systematically developed and a typical case and numerical consideration is given under regular or Poisson arrival and general service with equal mean. Finally, it is remarked that this original material is already written in Matsui (1982), and is recently developed to the cost factor case in Matsui (2008).

1. INTRODUCTION

A great many of the links in modern production systems (or processes) are formed by conveyors. In the area of conveyor theory, the two production systems: (i) mechanical (or moving-belt) flow-line system [38,pp.93-101] and (ii) open loop (or non-recirculating) conveyor system (e.g.[19][30][31]) are especially distinguished from a view of material-flow and studied in terms of the operational setting. These systems are called the *Conveyor-Serviced Production System* (CSPSystem) in the sense of mechanical material-flow system with variable arrival / service-times [22] [27]. Finally, it is noted that this original material is written in Matsui (1982), and is recently developed in Matsui (2008).

CSPSystem is a class of queueing networks with lost units, and is one of complex and intractable queueing systems. For example, the CSPSystem (ii) may be treated as a multi-channel queueing system with ordered-entry (e.g.[8][10]), but the distance between stations and the detailed activity of each operator are neglected. The CSPSystem (i) does not necessarily correspond to a queueing system of tandem type [12][19], since blocking does not occur in the sense that arriving units overflow when an operator is busy, and thus, is studied by industrial practice or computer simulation (e.g.[4][5][18]). Only non-mechanical (or non- paced) lines may be treated as the queueing system of tandem type (see [38,111-132]).

The traditional approach to CSPSystem is classified in the two types below. Such a treatment as the queueing system with ordered-entry or of tandem type iscalled the *system-centered approach*. The treatment that decomposes CSPSystem to each independent station (queueing sub-system) is called the *station-centered approach*. One of the queueing sub-system for the case of fixed items is a moving-operator without delay [16][21], while the Conveyor –Serviced Production Station (CSPS) is a typical queueing sub-system for the case of removal items with delay [2][25][26][28][29][33].

The station-centered approach is superior to the system-centered approach in terms of treating the detailed (or practical) model and designing each station buffer of CSPSystem, but it is lacking the basic theory that decomposes or synthesizes each queueing sub-system and designs the totally optimal CSPSystem. This reason results in the fact that the property of departure or overflow processes is not simple except the type of M/M/s [3][7][9], and thus, is an obstacle to the smooth development of conveyor theory.

Let us prepare a basic theory for the station-centered approach, and consider to design a CSPSystem by the mathematical programming with 2-level structure [11] [35]. The most important design variable of the production manager (decision maker) is the so-called cycle time of CSPSystem. In the area of stochastic line balancing concerning the CSPSystem (i), some papers try to approximately determine the minimum cycle time for a given number of stations [13][17]. Recently, a design procedure has beenproposed for directly determining the optimal cycle time under a given station buffer [32]. However, the cycle time problem of the CSPSystem (ii) would be unsolved, and the simultaneous determination of cycle time and station buffer (or time-range) does not seem to be done for both types [31].

This chapter is originated in [23] and distinguishes the CSPSystems, (i) and (ii), as Models I and II , respectively, composed of a decision maker and K CSPSs, and treats those as the two typical problems of queueing networks with the decision maker and lost units.

First, a station-centered approach to the class of CSPS Network is presented, and a few queueing formulas and general relation concerning average criteria are prepared. Next, a 2-level mathematical formulation of max-max and max-min types is presented for determining both the cycle time and time-range (control variable) of optimizing the production rate of Models and II ,respectively. In addition, a special case and numerical consideration is given under regular or Poisson arrival and general service with equal mean.

2. Two Models of CSP System

The CSPSystems, (i) and (ii), may be distinguished as Models I and II , respectively, composed of a decision maker and production processes (or line). The production processes of Model I are a series' array of K CSPSs (unloading and loading stations), while those of Model II are an ordered-entry array of K CSPSs (unloading stations). The relation between the production processes and decision maker is regarded as the two levels of hierarchy : CSPS units (1_{st} level) and a coordinator balancing CSPS units (2_{nd} level) (see Fig.1). An explanation of Models I and II is done here, and the 2-level mathematical structure will be determined in the later sections.

2.1. Material Flow and Unit

The mean input interval time is a design (or decision) variable of the coordinator, and is denoted by d ($0< d >\infty$). The *production rate*, r_i, i = 1, 2, ---, K, is defined as an inverse of the mean inter-departure time, i.e., mean time between successive departures. The *overflow rate*, v_i, i = 1, 2, ---, K, is defined as an inverse of mean inter-overflow time, i.e., mean time between successive overflows. In the Models I and II, the following relation is generally satisfied:

$$r_{i-1} = r_i + v_i, \text{ Model I}, \ i = 1, 2, ---, K \quad (2.1)$$
$$v_{i-1} = r_i + v_i, \text{ Model II}, \ i = 1, 2, ---, K$$

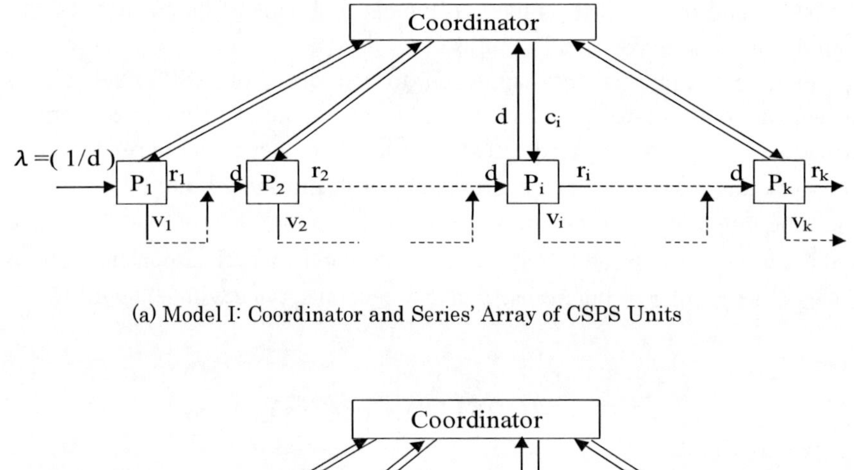

(a) Model I: Coordinator and Series' Array of CSPS Units

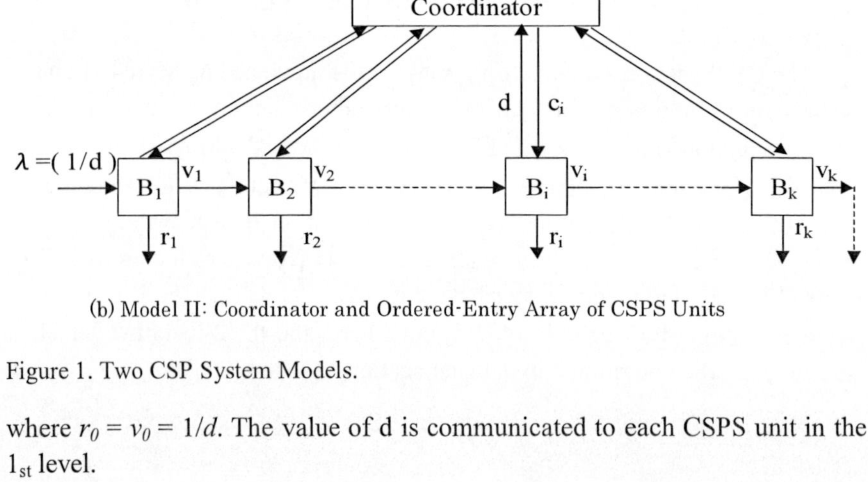

(b) Model II: Coordinator and Ordered-Entry Array of CSPS Units

Figure 1. Two CSP System Models.

where $r_0 = v_0 = 1/d$. The value of d is communicated to each CSPS unit in the 1_{st} level.

A practical assumption for CSPSystem is introduced to produce the production quantity required in the planning period. This is easy if it is able to approach the interdeparture time to the input interval time, and is realized by providing the large buffer within stations. Thus,

Assumption 1: $r_i + v_i = 1/d$, Model Ⅰ, $i = 1, 2, ---, K$

(2.2)

$\sum_{i=1}^{K} r_i + v_K = 1/d$, Model Ⅱ,

Under this assumption, the input interval time, d, is called the *cycle time*. An estimated value of d, T_c, is obtained from the planning period divided by the production quantity.

Each CSPS is manned by a single operator which obtains arriving usables in accordance with the operating policy, reserve-dependent and Sequential Range Policy (RdSRP). Figure 2 is the flowchart of the RdSRP for unloading stations, and shows a cycle of productive activity (called *work-cycle time*). The work-cycle time, Z, is a service time, X, plus a *delay* (or *idle*) *time* involved in obtaining usables from the conveyor. *Usables* are units suitable for utilization by the CSPS, and the usables that arrive during the service time overflow along the conveyor.

Arriving usables may be removed from the conveyor and stored in a storage facility with capacity N or N_i (called *reserve*). Work commences immediately on a usable in the reserve, and the processed unit is stored in another facility with infinite capacity (called *bank*). The number of holes contained in the reserve, n (= 1, 2, ---, N), is observed at the instant after the processed unit is stored in the bank. A *time-range* (control variable), c_n or c_n^i ($0 \leqq c_n < \infty$), is the time interval which the operator is able to look-ahead and observe arriving usables, and a control vector c_i is represented by $c_i = (c_1^i, c_2^i, ---, c_{N-1}^i)$. The value of c_i communicated to the coordinator in the 2_{nd} level.

The flowchart for unloading and loading stations is similar to Figure 2, except that a processed unit in the bank is placed on instead of obtaining a usable from the conveyor. The single Unit Policy (SUP) [30] is a special case of $N = 1$ or $c_1 = c_2 = --- = c_{N-1} = 0$, the unloading station with SUP is similar to the counter of type I [26], and the unloading and loading station with SUP may be regarded as a model of the moving-operator with delay [28]. The Sequential Range Policy (SRP) [2] [25] is equivalent to RdSRP under $c_1 = c_2 =$

--- = c_{N-1} = c. The CSPS with RdSRP would be distinguished as a qeueing-control model of look-ahead type (not seen in [6]). Recently, a summary of CSPS model is given in [24]. Then, the production and overflow rates are related to the *probability of processing*, P_i, and the *probability of loss*, B_i (= 1-P_i), respectively, which are a function of the control vector $\mathbf{c}_i$. That is ,

$$r_i = \lambda P_i = 1/Z_i = 1/(X_i + D) = 1/(1+H_i) = \lambda(1 - B_i), \quad (2.3)$$
$$v_i = \lambda - r_i = H_i/(1+H_i) = \lambda B_i$$

where a relation of D and H is linear and $\lambda D = 1 - \rho + H$ [24].

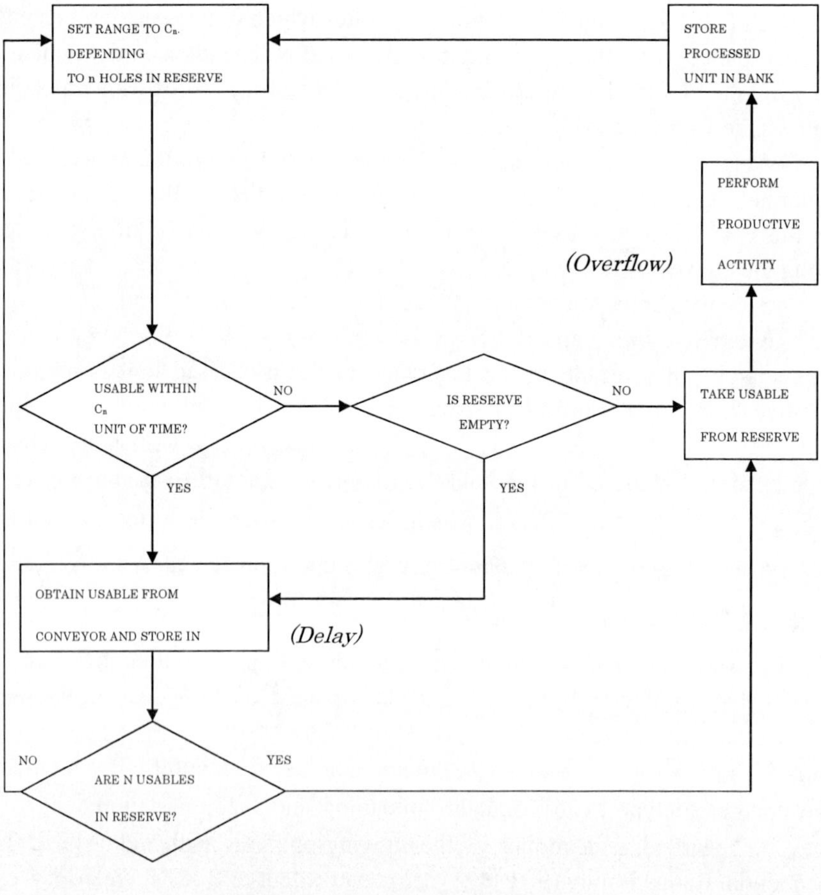

Figure 2. Optimal Operating Policy for Unloading Station.

2.3. Two Criteria for Models I and II

A general formula for the production rate of Models I and II is obtained from the fact that the formula (2.3) needs no specific assumption about the independence of inter-arrival interval. The production rate of Model I, r_I, is easily obtained from (2.2) and (2.3) as follows:

$$r_I = r_k = \lambda\prod_{i=1}^{K} P_i = \lambda\prod_{i=1}^{K}\{1/(1+H_i)\}, \tag{2.4}$$

in which H_i is the mean overflow number of units of the *i*th station. From (2.4), the overflow rate of Model I, v_I, is $v_I = \lambda - r_I$.

For Model II, the overflow rate of the *i*th station, v_i, becomes the inter-arrival rate of the (i+1)th station. From (2.2) and (2.3), the overflow rate of Model II, v_{II}, is easily given by

$$v_{II} = v_k = \lambda B_{II} = \lambda\prod_{i=1}^{K} P_i = \lambda\prod_{i=1}^{K}\{1/(1+H_i)\}, \tag{2.5}$$

Thus, the production rate of Model II, r_{II}, is from (2.5) as follows:

$$r_{II} = \lambda - v_{II} = \lambda(1 - \prod_{i=1}^{K} B_i) = \lambda[1 - \prod_{i=1}^{K}\{H_i/(1+H_i)\}], \tag{2.6}$$

If the probabilities P and B are regarded as a function of d(= 1/λ), the following property is obvious since the criterion H is a monotone decreasing function of d and ranges from zero to infinity. That is, Lemma 1. The probability of processing (loss), P(B), is non-negative (non-negative) for d ≧ 0, and is a monotone increasing (decreasing) function of d.

From (2.4) and (2.6), a simple inequality of production rates r_I and r_{II} are, respectively, as follows:

$$r_I|_{N=1} \leqq r_I \leqq r_I|_{N=1} \leqq \lambda, \qquad \text{Model I}$$

$$r_{II}|_{N=1} \leqq r_{II} \leqq r_{II}|_{N=1} \leqq \lambda, \qquad \text{Model II} \tag{2.7}$$

where the upper bound of r_I and r_{II} are respectively

$$r_I|_{N=\infty} = \lambda \min_i P_i|_{N=\infty}, \qquad r_{II}|_{N=\infty} = \lambda(1 - \min_i B_i|_{N=\infty}). \tag{2.8}$$

Also, a general formula for mean material-flow time is given below. Let W_i and L_i represent the mean material-flow time and time-average number of units in the *i*th station. Similar to other queueing formula, W = ZL needs no specific assumption about the independence of inter-arrival interval.

The mean material-flow time of Model I , F_I, is the mean throughput time for K stages, and is easily obtained from (2.2) and (2.3) as follows:

$$F_I = \sum_{i=1}^{K} W_i = \lambda^{-1} \sum_{i=1}^{K} L_i (1 + H_i), \tag{2.9}$$

$$F_{II} = (\textstyle\sum_{i=1}^{K} L_i) / r_{II}$$

$$= \lambda^{-1} (\textstyle\sum_{i=1}^{K} L_i) / [1 - \prod_{i=1}^{K} \{H_i/(1 + H_i)\}]. \tag{2.10}$$

Note that (2.4), (2.6), (2.9) and (2.10) consist of either only H_i's or H_i's and L_i's.

3. 2-Level Mathematical Structure

An optimization problem of the production rate of Models I and II is discussed, and the respective simultaneous determination of cycle time and time-range are regarded as the 2-level problem of maximizing the objective function of Models I and II , in which each CSPS may be pursuing its own objective. Under the Assumption 1, the objective functions R_I and R_{II}, moreover, are considered for the case of regular or Poisson arrival, general service with equal mean, and a property of the functions and their optimizations are discussed.

3.1. Objective Function of Coordinator

Now, let us consider the problem of optimizing the production rates r_I and r_{II} with respect to d(= 1/λ). Since the functions r_I and r_{II} have not necessarily the maximum value, a few discussions would be needed. The following lemma is primitives:

Lemma 2. Assume that $f(x) = x$, $x \geqq 0$, and

(i) $g(x), h(x) \geqq 0, x \geqq 0$

(ii) g is a monotone increasing function of x, and h is a monotone decreasing function of x.

Then, the function g/f and h·f have the maximum value, if

$$g = o(x) \text{ and } h = o(1/x), \tag{3.1}$$

respectively, when x approaches to zero and infinity.

Also, from (2.4) ~ (2.8), the coordinator would be valid to adopt the following objective function:

$$R_I(d, \mathbf{c}) = (1/d)P_I(d, \mathbf{c}) = (1/d) \prod_{i=1}^{K} P_i(d, c_i), \text{Model I} \tag{3.2}$$

$$R_{II}(d, \mathbf{c}) = dB_{II}(d, \mathbf{c}) = d \prod_{i=1}^{K} B_i(d, c_i), \text{ Model II} \tag{3.3}$$

where the extended vector c is c = (c_1, c_2, ---, c_{N-1}). The function R_I corresponds to the production rate, i.e., the rate that arriving units are processed throughput K stages, while the function R_{II} does to the busy time of Model II, i.e., the proportion that arriving units during the input interval time overflow from Model II.

From Lemmas 1 and 2, a property of the functions R_I and R_{II} is as follows:

Theorem. The functions R_I and R_{II} have the maximum value with respect to d, if the condition (3.1) is satisfied. Then, the function R_{II} has a saddle point, and the respective optimality conditions are

$$R_I(d, \mathbf{c}) \leqq R_I(d^\circ, \mathbf{c}^\circ), \tag{3.4}$$

$$R_{II}(d, \mathbf{c}^\circ) \leqq R_{II}(d^\circ, \mathbf{c}^\circ) \leqq R_{II}(d^\circ, \mathbf{c}). \tag{3.5}$$

Proof: The first is obvious, if x, g and h are replaced by

$$d, \prod_{i=1}^{K} P_i(d, c_i) \text{ and } \prod_{i=1}^{K} B_i(d, c_i)$$

in Lemma 2, respectively, and the Lemma 1 is applied. The inequalities (3.4) and (3.5) are the direct result of the first proof and the existence of optimal range strategy.

Remark 2: In the case of $K = 1$, the condition (3.1) would not be satisfied. For example, $g = P_1 = \mu d / (1+\mu d)$ and $h = B_1 = 1 / (1+\mu d)$ under Poisson arrival and

$N = 1$, and then, (3.1) does not hold. Thus, the function R_I and R_{II} have not the maximum value.

3.2. Special Case of Function R_I

Under the Assumption 1, a special case of the objective functions R_I and R_{II} is given and a property of optimization is discussed. First, let us consider a special case of the objective functions $R_I(d, c)$ with respect to the cycle time d.

For simplicity, the following is assumed instead of the Assumption 1 in Model I :

Assumption 2: At each stage of Model I ,the distribution of inter-arrival time is identical to that of input interval time.

For the case of $N_1 = N_2 = \cdots = N_K = 1$, the function $R_I(d, c)$ is given under the Assumption 3, and its optimization is done.

(i) Poisson arrival and general service; Since $H_i(d, c_i) = 1 / (\mu d)$ from $H_i = \rho$, the function $R_I(d, c)$ is obtained from (2.3) and (3.2) as follows:

$$\mu^{-1} R_I(d, c) = (\mu d)^{K-1} / (1 + \mu d)^K. \tag{3.6}$$

The optimal value $(\mu d)^*$ that maximizes (3.6) is given by $(\mu d)^* = K-1$. When $(\mu d)^*$

$= K-1$,

$$H_i(d, c_i) = 1 / (K-1), \sum_{i=1}^{K} B_i(d^*, c_i) = 1, K \geqq 2 \tag{3.7}$$

(ii) Regular arrival and Erlangian service with phase K; The function $R_I(d, c)$ is from (2.3), (3.2) and [26] as follows:

$$\mu^{-1} R_I(d, c) = (\mu d)^{-1}[1 + \sum_{m=0}^{k-1} (k\mu d)^m U_m(\exp(-k\mu d)) / m!]^{-K},$$
(3.8)

where $U_m(x)$, $0 < x < 1$, is given by

$$Um(x) = \sum_{u=1}^{\infty} umxu, \quad Um+1(x) = mU'm(x),\ m = 0, 1, 2, \text{---} \qquad (3.9)$$

Especially for the case of K = 2 and exponential service (k=1), (3.8) becomes

$$\mu^{-1} R_I(d, c) = \{1 - exp(-\mu d)\}^2 / (\mu d), \qquad (3.10)$$

and the optimal value $(\mu d)^*$ is given by

$$2 (\mu d)^* \{1 - exp(-(\mu d)^*\} = 1 - exp(-(\mu d)^*) \; or \; (\mu d)^* = 1.256.$$

(iii) Regular arrival and constant service; The following result is easily obtained for K = 2 (m = 1, 2, ---) :

$$H_i(d, c_i) = m-1,\ \mu^{-1} R_I(d, c) = (m^2 \mu d)^{-1},\ m-1 \leqq \mu d < (m-1)^{-1} \qquad (3.11)$$

From (3.11), it is obvious that $(\mu d)^* = 1$ and$\mu^{-1} R_I(d^*, c) = 1$. Generally, for any $K(\geqq 2)$, $(\mu d)^* = 1$ and $\mu^{-1} R_I(d^*, c) = 1$, and then

$$H_i(d^*, c_i) = 0, \sum_{i=1}^{K} B_i(d^*, c_i) = 0,\ K \geqq 2 \qquad (3.12)$$

From (3.7) and (3.12), a property of optimization may be found for Model I , and is as follows:

$$0 \leqq \sum_{i=1}^{K} B_i(d^*, c_i) \leqq 1, \text{Model I}, K \geqq 2. \qquad (3.13)$$

This property holds under the coefficient of variation that ranges from zero to 1, and shows that Model I satisfies the condition of the probability of loss.

3.3. Special Case of Function R_{II}

Next, a special case of the objective function $R_{II}(d, c)$ with respect to the cycle time d is considered. For simplicity, the following is assumed with the Assumption 1 in Model II:

Assumption 3. Under Poisson arrival, the distribution of inner-overflow time is approximated to the exponential type.

Under the Assumptions 1 and 2, the function $R_{II}(d, c)$ and its optimization are given for the case of $N_1 = N_2 = \cdots = N_K = 1$ or $N_1 = 1$, $N_2 = 2$. The validity of the Assumption 4 is obtained from a computer simulation [7] and the robustness for non-Poisson arrival [26]. In the case of K=3 and $N_1 + N_2 + N_3 = 5$, Table 1 shows

a comparison result of the assumption to the computer simulation for the production rate r_{II}.

(i) Poisson arrival and general service; For the case of $N_1 = N_2 = \cdots = N_K = 1$, the following relation for H_i's are obtained from (H=ρ) as follows:

$$H_i = H_{i-1}^2 / (1 + H_{i-1}), i = 2, 3, \cdots \quad (3.14)$$

where $H_1 = 1 / (\mu d)$. From (3.3) and (3.14), the function $R_{II}(d, c)$ is generally expressed as follows:

$$\mu R_{II}(d, c) = \mu d \prod_{i=1}^{K} B_i, i=2,3,\cdots \quad (3.15)$$

where $B_1 = H_1 / (1 + H_1) = 1/ (1+\mu d)$ and

$$B_i = B_{i-1}^2 / (1 - B_{i-1} + B_{i-1}^2), i = 2, 3, \cdots \quad (3.16)$$

Especially for the case of K = 2, (3.15) becomes

$$\mu R_{II}(d, c) = \mu d / \{1 + 2\mu d + 2(\mu d)^2 + (\mu d)^3\}, \quad (3.17)$$

and the optimal value $(\mu d)^*$ is given by $(\mu d)^{*2} + (\mu d)^{*3} = 1/2$, i. e., $(\mu d)^* = 0.565$. Then, $\mu R_{II}(d, c) = 0.192$.

Remark 3. For the same case of the Erlang's loss formula E_K, the corresponding result is given by

$$\mu dE_2 = \mu d \,/\, \{1 + 2\mu d + 2(\mu d)^2\} \,, \tag{3.18}$$

and the optimal values are $(\mu d)^* = 1/\sqrt{2} = 0.707$ and $\mu dE_2^{'} = 0.207$. This result shows that the optimal cycle time would exist in a multi-channel queueing system with ordered- entry.

In the case of N_1=1 and N_2=2, the function $R_{II}(d, c^*)$ is from (3.3), $H_1 = 1/(\mu d)$ and $H_2 = \rho\{1 - (1 - P_1)S_0(c^*)\}$ and (3.3) as follows:

$$\mu R_{II}(d, c^*) = \frac{\mu d}{1+\mu d}\,\frac{1-X}{1+\mu d+(\mu d)^2 - X} \tag{3.19}$$

Table 1. Approximation Error of Production Rate r_{II}

d=1, μ=0.5 and $N_1+N_2+N_3=5$							
	service	exponential		Erlangian(k=4)		constant	
reserve		App.	Sim.	App.	Sim.	App.	Sim.
$N_1=N_2=1$	r_{II}	0.8489	0.8875	0.8314	0.8517	0.8425	0.8660
$N_3=3$	relative error (%)	4.3		2.4		2.7	
$N_1=N_3=1$	r_{II}	0.8521	0.8535	0.8360	0.8304	0.8475	0.8395
$N_2=3$	relative error (%)	0.2		-0.7		-1.0	
$N_1=1$	r_{II}	0.8749	0.8895	0.8503	0.8530	0.8645	0.8754
$N_2=N_3=2$	relative error (%)	1.6		0.3		1.2	

where the extended vector c* is c= (0, c*) and X is

$$X = S_0(c^*)\exp\{-c^*/[\mu d/(1+\mu d)]\},\ \lambda S_0(c^*) = S_0^{'}(c^*). \tag{3.20}$$

(ii)Regular arrival and constant service; For the case of $N_1 = N_2 = 1$, the function $R_{II}(d, c)$ is easily given by

$$\mu R_{II}(d,c) = \begin{cases} 0, & d \geqq 1/2 \\ \mu d(m-2)/m, & m^{-1} \leqq \mu d < (m-1)^{-1},\ m = 3,4,---. \end{cases} \tag{3.21}$$

and the optimal values are $(\mu d)^* = 1/2$ and $\mu R_{II}(d^*, c) = 0$. Generally, for any $K(\geqq 2)$, the optimal values are $(\mu d)^* = 1/K$ and $\mu R_{II}(d^*, c) = 0$, and then,

$$H_i(d^*, c) = K - 1, \Sigma_{i=1}^{K} P_i(d^*, c) = 1, K \geqq 2. \quad (3.22)$$

4. Formulation and Numerical Consideration

A 2-level mathematical formulation of max-max and max-min types is presented for determining both the cycle time and time-range of optimizing the objective function of Models I and II, respectively, under the constraint of mean material-flow time. A numerical example is given under the Assumptions 1~4, regular or Poisson arrival and general service with equal mean, and a few properties of optimization are numerically discussed. Especially, an interesting comparison of the two material-flow system: series vs. ordered-entry given.

4.1. 2-Level Mathematical Consideration

A basic coordination problem of Models I and II without cost factor or tradeoff is to maximize the objective functions R_I and R_{II} of Models I and II, in which each CSPS pursues the maximization of the probability of processing, P_i, and minimization of the probability of loss, B_i, respectively, with respect to the control vector c_i. Thus, the two basic problems under the constraint of mean material-flow time may be formularized by 2-level mathematical programming as follows:

Basic Problem I $max_d (1/d) \prod_{i=1}^{K} P_i(d, \hat{c}_i(d))$ (4.1a)

s.t. $0 < d \leqq T_c$, (4.1b)

$\sum_{i=1}^{K} B_i(d, \hat{c}_i(d)) \leqq \alpha_I$ (4.1c)

$$\sum_{i=1}^{K} W_i(d, \hat{c}_i(d)) \leqq \alpha_2 \quad , \tag{4.1d}$$

$$P_i(d, \hat{c}_i(d)) = max_{\hat{c}i} \{1/[1+H_i(d, c_i)], \tag{4.1e}$$

$$\text{s.t. } 0 \leqq c_i \leqq c_0^i, \tag{4.1f}$$

$i = 1, 2, ---, K$

$$\text{Basic Problem II}: max_d\, dB_{II}(d, \hat{c}(d)) \tag{4.2a}$$

s.t. $0 < d \leqq T_c$, (4.2b)

$$\sum_{i=1}^{K} P_i(d, \hat{c}_i(d)) \geqq \beta_1, \tag{4.2c}$$

$$F_{II}(d, \hat{c}(d)) \leqq \beta 2, \tag{4.2d}$$

$$B_{II}(d, \hat{c}(d)) = \min_c \prod_{i-1}^{K} H_i(d, c_i)/[1+H_i(d, c_i)]\} \ , \tag{4.2e}$$

$$\text{s.t. } 0 \leqq c \leqq c_0, \tag{4.2f}$$

respectively, in which the constants $\alpha_1, \beta_1, \alpha_2$ and β_2 are $0 \leqq \alpha_1, \beta_1 \leqq 1$ and $0 < \alpha_2, \beta_2 < \infty$, and the vectors c_0^i and c_0 are constant.

If all functions of Basic Problems I and II are convex of concave, a few methods for solving the Basic Problem I are seen in [1] and [11], while the Basic Problem II would be solved by a few methods of the max-min problem with 2-level structure (see[35]). In the sense that the program package of SUMT is available, a penalty method [1] [36] would be useful. Thus, a numerical example below is limited to a simple case of Basic Problems I and II.

A numerical result of the functions R_I and R_{II} is given from the result above, and a simple example of Basic Problems I and II is numerically solved. With respect to the cycle time d, Figures 3 and 4 show the behavior of the functions R_I and R_{II}, respectively, in the case of K=2. Figure 5 shows the behavior of the optimal value $(\mu d)^*$ in the case of K = 2 (1) 10. Tables 2 and 3 give the corresponding values to Figure5 in Models I and II, respectively. Table 4 shows the optimal solution of the Basic Problem I, (6.1a) and (6.1e), and Basic Problem II, (6.2a) and (6.2e), in the case of $N_1=N_2=2$, Poisson arrival and Erlangian service with phase k. The busy probability, 0_i, is given by $0_i = 1/(\mu Z_i)$ or $= 1/\ \{\mu d(1+H_i)\}$.

Viewing Figsures 3 and 4, it is ascertained that the functions R_I and R_{II} are uni-model and concave (or pseudo-concave) except the neighborhood of constant service (k=∞). From Figure 5 , it is seen that, needless to say, the optimal value (μd)* of Models I and II are a monotone increasing and decreasing function of K, respectively. Tables 2 and 3 indicates that each busy probability of CSPSs in Models I and II are decreasing and increasing to according to the increase of K, respectively. From Figures 3 and 5, it is noted that the optimal value (μd)* of K=2 increases and decreases according to the increase of Erlangian phase k.

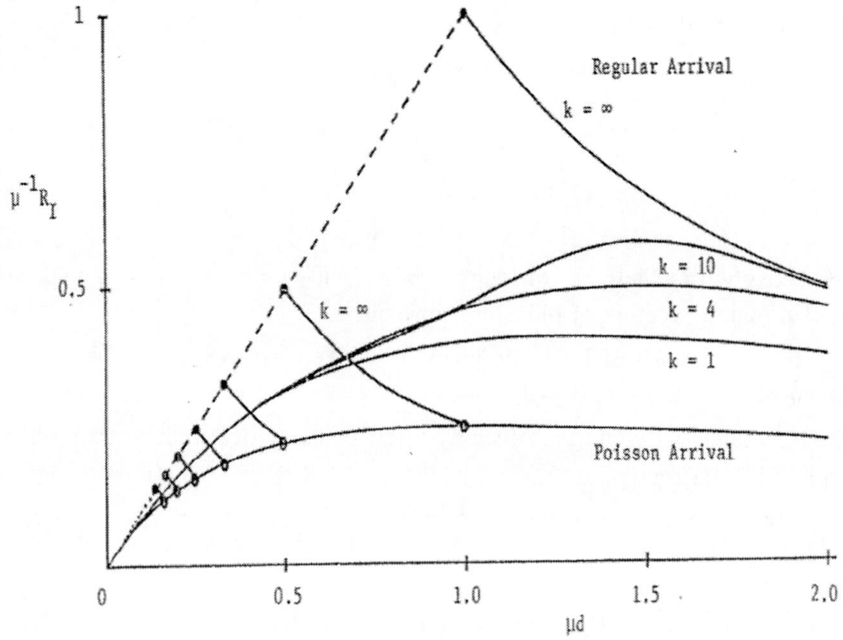

Figure 3. Case of K=2: Model I, $N_1=N_2=1$, Poisson Arrival and General Service, Regular Arrival and Erlang Service (phase K).

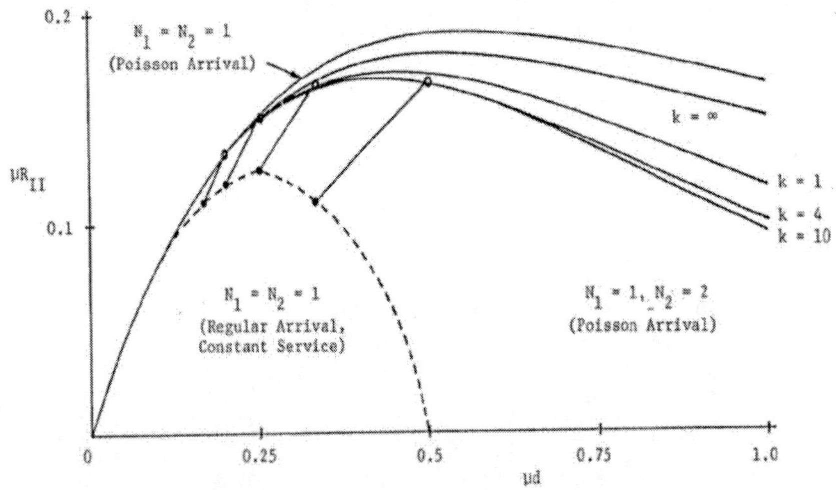

Figure 4. Case of K=2: Model II, N_1=N_2=1, 2(c=d), Poisson Arrival and General Service, Regular Arrival and Erlang Service (phase K).

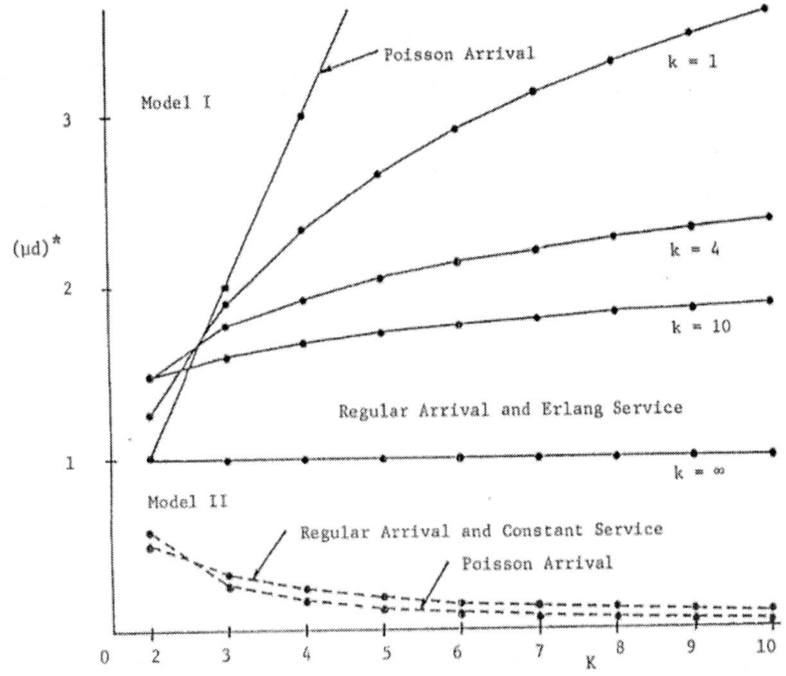

Figure 5. Behaviour of the Optimal Value (ud)*: K=2 (1) 10, N_1=N_2 = --- = N_K=1.

Table 2. Optimal Values of Model Ⅰ : K = 2, 3, 5, 10, Poisson arrival and general services, Regular arrival and Erlang service (phase k)

type \ model K		Model I			CSPS	
		$(\mu d)^*$	$\mu^{-1}R^*$	$(\mu d)^{-1}P_K$	P_i	O_i
M/G	2	1.000	0.250	0.500	0.500	0.500
	3	2.000	0.148	0.333	0.667	0.333
	5	4.000	0.082	0.200	0.800	0.200
	10	9.000	0.039	0.100	0.900	0.100
k=1	2	1.256	0.407	0.569	0.715	0.569
	3	1.904	0.324	0.447	0.851	0.447
	5	2.660	0.262	0.350	0.930	0.350
	10	3.615	0.211	0.269	0.973	0.269
k=4	2	1.469	0.501	0.584	0.858	0.584
	3	1.766	0.451	0.525	0.927	0.525
	5	2.051	0.407	0.470	0.964	0.470
	10	2.376	0.363	0.415	0.985	0.415
k=10	2	1.468	0.583	0.630	0.925	0.630
	3	1.594	0.550	0.601	0.957	0.601
	5	1.728	0.518	0.566	0.978	0.566
	10	1.886	0.482	0.525	0.991	0.525

Table 3. Optimal Values of Model Ⅱ : K = 2, 3, 5, Poisson arrival and general service

model	K	2		3		5	
		P_i	B_i,O_i	P_i	B_i,O_i	P_i	B_i,O_i
CSPS	i = 1	0.361	0.639	0.212	0.788	0.115	0.885
	2	0.469	0.531	0.254	0.746	0.128	0.872
	3	-	-	0.314	0.686	0.144	0.856
	4	-	-	-	-	0.164	0.836
	5	-	-	-	-	0.191	0.809
Model II	$(\mu d)^*$	0.565		0.269		0.130	
	μR^*_{II}	0.192		0.108		0.085	
	$\mu^{-1}r_{II}$	1.17		2.22		4.26	

4.2. Optimization Properties

The following is seen from Table 3:

$$P_i(d^*, c_i) \leqq 1/K, 1/2 \leqq \sum_{i=1}^{K} P_i(d^*, c_i) \leqq 1, K \geqq 2. \quad (4.3)$$

From (3.22) and (4.3), the following property of optimization may be found for

Model II :

$$1/2 \leqq \sum_{i=1}^{K} P_i(d^*, c_i) \leqq 1, \text{Model II}, K \geqq 2. \quad (4.4)$$

Thus, a property of the optimization of Models I and II are given by (3.13) and (4.4), respectively, and would hold under a more general type of arrival. From this property, the constraints (4.1c) and (4.2c) are imposed on the Basic Problems I and II , respectively.

From Table 4, the effect of reserve is easily ascertained, and the property (3.13) or (4.4) is confirmed. Also, the optimal time-range of Models I and II with equal service time satisfy the respective following relations: $c^1* = c^2*$, and $c^1* \leqq c^2*$, and thus, the respective relations may be generalized for K $\geqq$2 as follows:

$$c^{1*} = c^{2*} = \text{---} = c^{K*}, \text{Model I} \quad (4.5)$$

$$c^{1*} \leqq c^{2*} \leqq \text{---} \leqq c^{K*}, \text{Model II} \quad (4.6)$$

in which the c^{1*} is optimal time-range of the *i*th CSPS. The former (4.5) is the symmetrity of Model I , while the latter (4.6) is the monotonicity of Model II .

Finally, an interesting comparison of the two material-flow systems: series and ordered-entry is given under $\Sigma(1/\mu)$= K and $1/\mu$= K, respectively. If the regular arrival and constant service with synchronization are assumed, no significant difference would be found in the both types. But, if the Poisson arrival and general service is assumed, the difference of both types is seen from Table 5.

Table 4. Case of $N_1 = N_2 = 2$: Poisson arrival and Erlang service (phase k)

model	level	2nd level			1st level (CSPS)			
					1		2	
Model I	k	$(\mu d)^*$	$\mu^{-1}R^*_I$	$(\mu d)^{-1}P_2$	c^{1*}	O_1	c^{2*}	O_2
	1	1.093	0.327	0.547	0.739	0.547	0.739	0.547
	2	1.113	0.343	0.555	0.759	0.555	0.759	0.555
	4	1.135	0.355	0.559	0.790	0.559	0.790	0.559
	10	1.166	0.366	0.560	0.836	0.560	0.836	0.560
Model II	k	$(\mu d)^*$	μR^*_{II}	$\mu^{-1}r_{II}$	c^{1*}	O_1	c^{2*}	O_2
	1	0.402	0.162	1.487	0.339	0.782	0.463	0.704
	2	0.385	0.159	1.527	0.347	0.798	0.467	0.728
	4	0.375	0.157	1.549	0.359	0.807	0.485	0.741
	10	0.369	0.157	1.560	0.368	0.810	0.512	0.749

Table 5. A comparison of Series and Ordered-Entry System:Poisson arrival, K = 2 (N = 1, 2, k = 10), K = 3, 5 (N = 1)

model	K	2		3	5
		N = 1	N = 2	N = 1	N = 1
Model I (series)	$1/\mu$	1.000	1.000	1.000	1.000
	d^*	1.000	1.166	2.000	4.000
	$(\mu d)^*$	1.000	1.166	2.000	4.000
	$(1/d^*)P_K$	0.500	0.654	0.333	0.200
	$(1/d^*)\Sigma B_i$	1.000	0.407	0.500	0.250
	ratio	2.000	0.622	1.500	1.250
Model II (ordered-entry)	$1/\mu$	2.000	2.000	3.000	5.000
	d^*	1.130	0.738	0.807	0.650
	$(\mu d)^*$	0.565	0.369	0.269	0.130
	r_{II}	0.585	0.780	0.740	0.852
	v_{II}	0.300	0.575	0.500	0.686
	v_{II}/r_{II}	0.513	0.737	0.676	0.805

For the case of $N_1 = N_2 = \text{---} = N_K = 1$, Table 5 shows that the system of ordered-entry type is superior to that of series type in terms of production rate. According to the increase of phase k and reserve capacity N, the latter would become superior to the former in terms of total overflow rate.

CONCLUSION

This chapter treats the two types of CSPSystems as the respective queueing networks composed of a decision maker (coordinator) and the production processes: a series' array (Model I) and an ordered-entry array (Model II) of K CSPSs. A few queueing formulas and general relation concerning average criteria are showed, and a mathematical theory (or method) is presented to analyze and design the CSPS Network on the base of station-centered approach.

Recently, a design approach by cost/profit factors is studied since 2001, and a lot of assembly-system cases are seen in Matsui (2008). Further researches will be hoped to treat the alternative CSPS-type unit and to challenge a dual problem of Model I and II in the conveyor-like network.

REFERENCES

Aiyoshi, E., & Shimizu, K. (1981), Hierarchical Decentralized System and its New Solution by a Barrier Method, IEEE Transactions SMC-11, 6, 444-449.

Beightler, C. S., & Crisp, R. M., Jr. (1968), A Discrete-Time Queueing Analysis of Conveyor-Serviced Production Stations, Operations Research 16, 5, 986-1001.

Burke, P. J. (1956), The Output of a Queueing System, Operations Research 4, 699-704.

Buxey, G. (1978), Incompletion Costs versus Labor Efficiency on the Fixed-Item Moving Belt Flowline, International Journal of Production Research 16, 3, 233-247.

Buxey, G. M., & Sadjadi, D. (1976), Simulation Studies of Conveyor-Paced Assembly Lines with Buffer Capacity, International Journal of Production Research 14, 5, 607-624.

Crabill, T. B., Gross, D., & Magazine, M. J. (1977), A Classified Bibliography of Research on Optimal Design and Control of Queues, Operations Research 25, 2, 219-232.

Crisp, R. M., Jr., Skeith, R. W., & Barnes, J. W. (1969), A Simulated Study of Conveyor-Serviced Production Stations, International Journal of Production Research 7, 4, 301-309.

Disney, R. L. (1963), Some Multi-channel Queueing Problems with Ordered Entry—An Application to Conveyor Theory, Journal of Industrial Engineering 14, 2, 105-108.

Disney, R. L. (1975), Random Flow in Queueing Networks: A Review and Critique, AIIE TRANSATIONS 7, 3, 268-288.

Erlang, A. K. (1917), Solution of some Probability problems of Significance for Automatic Telephone Exchanges, Post Office Electrical Engineer's Journal 10, 189-197.

Geoffrion, A. M., & Hogan, W. W. (1972), Coordination of Two-Level Organization with Multiple Objectives, In A. Y. Balakrishnan (Ed.), Techniques of Optimization New York: Academic Press, 455-468.

Hunt, G. C. (1956), Sequential Arrays of Waiting Lines, Operations Research 4, 6, 674-683.

Ignall, E. J. (1965), A Review of Assembly Line Balancing, Journal of Industrial Engineering 16, 4, 244-254.

Jewell, W. S. (1967), A Simple Proof of: $L = \lambda W$, Operations Research 15, 6, 1109-1116.

Kottas, J. F., & Lau, H. S. (1981), A Stochastic Line Balancing Procedure, International Journal of Production Research 19, 2, 177-193.

Kuroda, T. (1968), A Study on the design of Production Systems in production control, Doctoral Dissertation, Tokyo: Waseda University (in Japanese).

Kwo, T. T. (1958), A Theory of Conveyors, Management Science 6, 1, 51-57.

Little, J. D. C. (1961), A Proof for the Queueing Formula: $L = \lambda W$, Operations Research 9, 3, 383-387.

Matsui, M. (1981), A Study on Optimal Operating Policies in Conveyor-Serviced Production System, Doctoral Dissertation, Tokyo: Tokyo Institute of Technology (in Japanese).

Matsui, M. (1982), Conveyor-Serviced Production System: Queueing Formulas, CSPS Analysis 2-level Mathematical Formulation, unpublished paper.

Matsui, M. (2005), CSPS Model: Look-Ahead Controls and Physics, International Journal of Production Research 43, 10, 2001-2025.

Matsui, M. (2008), Manufacturing and Service Enterprise with Risks: A Stochasitic Management approach, New York: Springer.

Matsui, M., & Fukuta, J. (1975), A Queueing Analysis of Conveyor-Serviced Production Station with General Unit-Arrival, Journal of the Operations Research Society of Japan 18, 3・4, 211-227.

Matsui, M., & Shingu, T. (1978), A Queueing Analysis of Conveyor-Serviced Production Station and the Optimal Range Strategy, AIIE TRANSATIONS 10, 1, 89-99.

Matsui, M., Shingu, T., & Makabe, H. (1977), Conveyor-Serviced Production System: An Analytic Framework for Station-Centered Approach by Some Queueing Formulas, Preliminary Report of the Operations Research Society of Japan, Autumn, 104-107 (in Japanese).

Matsui, M., Shingu, T., & Makabe, H. (1978), An Analysis Conveyor-Serviced Production Station by Queueing Theory, Journal of Japan Industrial Management Association 28, 4, 375-286 (in Japanese).

Matsui, M., Shingu, T., & Makabe, H. (1980), A Comparative Consideration of Operating Policies for Conveyor-Serviced Production Station, Journal of Japan Industrial Management Association 31, 3, 342-348 (in Japanese).

Morris, W. T. (1962), An Analysis for Materials Handling Management, Illinois: Richard D. Irwin. Inc., 129-169.

Muth, E. J., & White, J. A. (1979), Conveyor Theory: A survey, AIIE TRANSACTIONS 11, 4, 270-277.

Nishiyama, N., & Nagasawa, H. (1980), A Design of Assembly Line with Distributed Operation Time (1st Report, Optimal Cycle-Time for Maximizing the Expected Production Volume under the Float-Type Buffer Storage)," Transactions of the Japan Society of Mechanical engineers, C-46, 404, 450-457: and other three papers (in Japanese).

Ries, L. L., Brennan, J. J., & Crisp, R. M., Jr. (1967), A Markovian Analysis for Delay at Conveyor-Serviced Production Station, International Journal of Production Research 5, 3, 201-211.

Ross, S. M. (1970), Applied Probability Models with Optimization Applications, San Francisco: Holden-Day.

Shimizu, K. (1981), A Solution to Mathematical Programming Problems with Two-Level Structure, Proceedings of the 2nd Mathematical Programming Symposium, Japan, 71-80.

Shimizu, K., & Aiyoshi, E. (1981), A New Computational Method for Stackelberg and Min-Max Problems by Use of a Penalty Method, IEEE Transactions AC-26, 2, 460-466.

Stidham, S., Jr. (1974), A Last Word on $L = \lambda W$, Operations Research 22, 2, 417-421.

Wild, R. (1972), Mass-production management, The Design and operation of Production Flow-Line Systems, London: John Wiley.

In: Manufacturing Engineering
Editors: Anthony B. Savarese
ISBN: 978-1-61209-987-3

Chapter 4

MANUFACTURING OF BRAKE FRICTION MATERIALS

Dragan Aleksendrić
University of Belgrade Faculty of Mechanical Engineering, Belgrade, Serbia

ABSTRACT

Process of brake friction material manufacturing has a crucial impact on the future properties of brake friction material. It is especially related to the level and stability of friction and wear during braking. Development of brake friction material is strongly affected by its formulation and manufacturing conditions. Due to complex and interrelated influences between formulation and manufacturing conditions of brake friction material, it is difficult to find out the best set of manufacturing conditions, for the specific formulation of friction material, which satisfy wanted friction and wear properties. In this chapter, influence of manufacturing conditions of the brake friction material has been investigated versus its wear. In order to investigate the character of influence of manufacturing conditions, effect of manufacturing parameters on wear properties of the brake friction material has been modelled. This model was able to recognize the way of how moulding pressure, moulding time, moulding temperatures, heat treatment time, and heat treatment temperatures affecting wear of the brake friction material. The model of influence of manufacturing parameters on wear of the brake friction material was based on artificial intelligence. The model is able to make a functional relationship between a formulation of friction material and effects of chosen manufacturing parameters on its wear. It provided possibilities for searching the best set of manufacturing conditions and their adaptation to a formulation of brake friction material.

Keywords: brake friction material, manufacturing conditions

1. INTRODUCTION

Friction materials are important parts in braking systems. They convert the kinetic energy of the car to thermal energy by friction in the contact zone. The complicated series of events that occur in the contact zone play a crucial role in the tribological behaviour of the brake lining materials [1]. Friction material is developed to meet comprehensive tribological and mechanical requirements. It is a mix of more than 20 different ingredients. The friction material in the automotive brake system could be taken as one of the key component responsible for overall functional characteristics of motor vehicle brakes. This is because it plays crucial roles in various aspects of the brake performance such as stopping distance, pedal feel, wear, and brake induced vibrations [2]. Competitive advantages in the industry of friction materials need to be reached by an appropriate management of friction material formulation and manufacturing conditions changing and their skilful implementation in a cost effective manner. Management of these changes can have decisive influence on whether a new friction material can be launched into the market. It is important because the industry of friction material manufacturing has been always focused on greater customer satisfaction i.e. (i) improved friction stability, (ii) improved life (iii) no judder, (iv) no noise, (v) improved rotor compatibility [3]. To understand the development of friction materials, it is necessary to understand the interfacial relations between rubbing surface. There are many asperities between the rubbing surfaces because the formation of many micro-contact points, these points are dynamically changing from place to place in fraction of seconds during rubbing [4,5,6,7]. The total contact area is unknown and depends on friction pair interaction. Furthermore, it is known that when two bodies slide against each other with a relative speed and a contact pressure, frictional heat is generated at their sliding interface [5]. The subsequent thermo-mechanical deformation of the bodies modifies the contact profile and the pressure distribution, altering the frictional heat. The friction heat generated during braking application easily raises the temperature at the friction interface beyond the glass transition temperature of the binder resin and often rises above decomposition temperature [8,9]. The gas evolution at the braking interfaces because of pyrolysis and thermal degradation of the material results in the friction force decreasing at elevated temperatures. The thermal load in the contact of brake pads is not built up homogeneously and hot rings can be often seen on the brake disc changing their radius. The instability of the friction coefficient after a certain number of brake applications is common and depends on contact interface situation. In general, the change in friction and wear of friction material during sliding depends on the changes of the real area of contacts at the friction interface, the strength of the binder resin, the frictional properties of ingredients at elevated temperatures, and friction film formation. Temperature sensitivity of brake friction materials has always been a critical aspect while ensuring their smooth and reliable functioning. It is particularly related to front brakes that absorb a major amount (up to 80%) of the vehicle total kinetic energy. The severity of such temperature rise is further manifested in the form of a very high flash temperature at the contacting asperities (600-800°C). At such high temperatures, friction force suffers from a loss of effectiveness called μ-fade [9] but also wear properties could be drastically changed. This loss of effectiveness and wear change of brake friction materials cannot be easily predicted due to different ingredients used for friction material's mixture, its different formulations, and different setting of manufacturing parameters. It is known that organic constituents, fibrous materials, and solid lubricants play important role in establishing the transfer layer at the friction interface which affecting both friction and wear. The clear connection between component mixtures used manufacturing conditions, friction layer, and the connection between friction layer and friction behaviour of the system is not known yet [9m]. There is no doubt that measures towards optimization of the friction materials versus their friction and wear properties have to be taken but trial and error method needs to be eliminated due to cost and time consuming. Accordingly, intelligent methods should be introduced in the process of brake friction material development. It is very important because selection of kind and concentration of ingredients in the brake friction material formulation and adaptation of manufacturing conditions to the specific formulation of brake friction material is needed. These influences strongly affected a contact situation between a disc and brake pad and consequently friction and wear characteristics during braking.

Manufacturing of brake friction materials is complex process because powder mixes, containing metal chips, filling agents, abrasives, and phenolic resins, are moulded to a back plate under pressure and temperature [10]. Such moulded friction material is further exposed to additional thermal treatment. This production process leads to anisotropic, visoelastic, and to certain extent to heterogeneous material [10]. The moulding and heat treatment parameters of the brake friction materials could be often chosen based on resin properties specified by resin manufacturer. Manufacturing conditions of brake linings are affected mostly by the thermal properties of the binder resin, such as flow distance and gel or 'B' times. Phenolic resins (modified and unmodified) are invariably used as binders in friction materials due to their low cost and good combination of mechanical properties, such as high hardness, compressive strength, moderate thermal resistance, creep resistance and very good wetting capability with most ingredients [1]. Furthermore, concentration of phenolic resin in the friction material should be reconciled with the share of others ingredients in the formulation of brake friction material. Not only that, manufacturing parameters such as moulding pressure, moulding time, and moulding temperature need to be properly set to correspond the formulation of friction material. Decisions made in this stage strongly affected the future friction and wear behaviour of brake friction materials. That is why, it is very important to know effects of different combination parameters related to formulation and manufacturing.

Wear in general relies on many factors such as temperature, applied load, sliding velocity, properties of mating materials, and durability of the transfer layer. Until now, a much effort has been paid to understand the wear mechanisms and develop a model for prediction of wear properties of different engineering materials [11,12,13,14,15]. Wear mechanisms operating in brake friction materials were found to be a complex combination of abrasion, adhesion, fatigue, delamination, and thermal decomposition [16]. That is why, stochastic nature of wear of brake friction materials is result of a dominant wear mechanism and its transition to another. The dominant wear mechanism of brake friction materials is influenced by braking times, applied loads, and material characteristics. Furthermore, the material characteristics are strongly affected by formulation and manufacturing conditions. Accordingly, prediction and particularly optimization of wear performance of friction materials is very complex task. The main problem is related to impossibility to correlate effects of many influencing factors against wear of the friction materials under different operating conditions of automotive brakes. It is particularly difficult if the composition of friction material and its manufacturing conditions are taken into consideration. The traditional mathematical techniques related to modelling of wear of brake friction material, explained in [17,18,19,20], were not able to provide the wear model with inherent abilities to generalize the complex synergy of different influences of friction material formulation, manufacturing, and a brake load conditions. An artificial intelligence, based on artificial neural networks, are suited for simulating complex polymer composite problems because it can be learned and trained to find solutions [21]. In this chapter, artificial intelligence has been used for mining of experimental data in order to model and predict the influence of manufacturing parameters of the brake friction material, which formulation (phenolic resin, fibers, abrasives, and lubricants) is known, on its wear.

2. EXPERIMENT

As it is mentioned, the brake's performance is primarily influenced by contact situation between a cast iron brake disc and friction materials. On the other hand, the contact situation is differently affected by wide diversity in mechanical properties of the friction material ingredients [22]. These ingredients can be grouped into at least four classes: (i) fibrous reinforcement, (ii) binder, (iii) filler and (iv) friction modifier. The diversity of chemical and mechanical properties of friction material ingredients, in some extent, is further changed by manufacturing parameters. The situation is more complicated if synergistic influences between these parameters are taken into consideration. Synergetic effects of all these ingredients, as is shown in Fig. 1, determine the final friction material characteristics.

It is evident that the brake performance results from the complex interrelated triboprocesses occurring during braking in the contact of the friction pair. These complex processes and the contact characteristics are mostly affected by the physicochemical properties of the friction materials ingredients. Moreover, the same friction material formulation can be differently affected by setting of manufacturing conditions defined by dry mixing, pre-forming, hot moulding, and heat treatment (post curing).

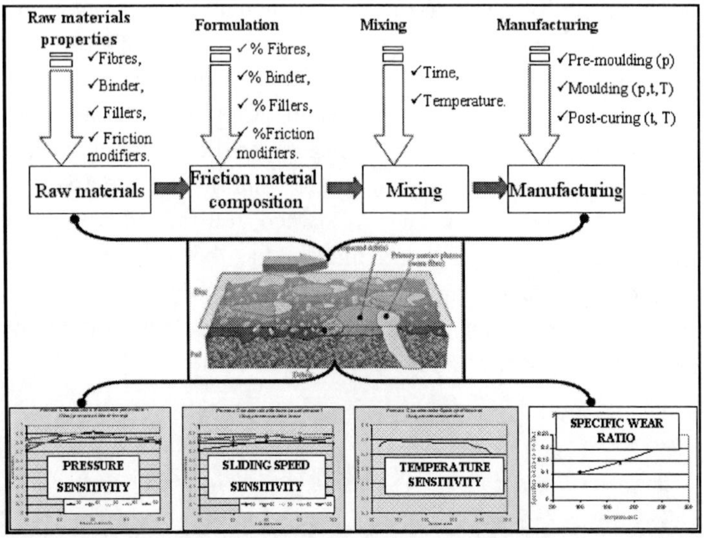

Figure 1. Synergetic influence of formulation and manufacturing conditions on friction and wear properties of brake friction material.

According to [1], the moulding involves several stages: initial moulding, elastic–plastic deformation, and particle fracture or fragmentation. The initial stage is mostly affected by the particle size and shape. As the applied pressure is further, increased, plastic deformation occurs locally at the interparticle contact points and here the mechanical properties and the quality of the particles are important factors, because they control the compressibility behaviour of the powder [1]. The consequences of different stages in the friction materials production and their interrelated influences are illustrated in Fig. 1 [22]. From Fig. 1, it can be seen that the most complicated task is related to establishing a relationship between friction materials formulation and manufacturing parameters versus imposed requirements related to friction and wear. This is particularly difficult taking into consideration complexity of following requirements imposed to the brake friction materials: (i) sensitivity of the friction coefficient to the line pressure application and/or sliding speed and/or temperature change, (ii) wear characteristics under different temperatures operating regimes, (iii) vehicle weight, and (iv) braking systems characteristics [22].

An experiment related to influence of manufacturing conditions of brake friction material on its wear properties should be carefully planed. It should be based on testing of different types of friction material. As explained so far and illustrated in Fig. 1, there is no doubt that conditions for friction material development are extremely difficult and soft-computing techniques have to be employed in order to solve this complex problem. Application of soft computing technique, such as artificial neural networks, demands clear identification of the model's input/output. In this chapter, the input parameters are defined by the friction material formulation i.e. mass percentages of phenolic resin, fibers, abrasives, and lubricants, manufacturing conditions of friction materials, and the disc brake operating conditions represented by work done by the brake applications and brake interface temperature. The friction material wear volume (mm^3) has been taken as an output parameter obtained after certain number of the disc brake applications i.e. after certain work done by the disc brake. The type of data generator depends on the application and the availability. In this case, as a data generator a single-end full-scale inertial dynamometer has been used, see Fig. 2.

The electromotor (1) drives via carrier (2) a set of 6 flywheels (3) located in protective cage (4), providing in such way different inertial masses. At the end of the common shaft is firmly jointed disc (5) of the tested brake. The stationary part of the tested brake – caliper (6) is connected with the system for braking torque measurement (7). Position of the system can be adjusted by axial slider (8). As it can be seen in Fig. 2, all these components are mounted on the common foundation (9). Thus, manufacturing conditions were investigated versus wear of the friction material, a testing methodology needs to be chosen according to the range and distribution of data that are going to be collected (table 1). A wear test has to be properly designed in order to provide the brake's average operation conditions. As it can be seen in table 1, a test is designed on a way that brake's application pressure and initial speed have been kept constant (initial speed of 90 km/h; application pressure of 40 bar).

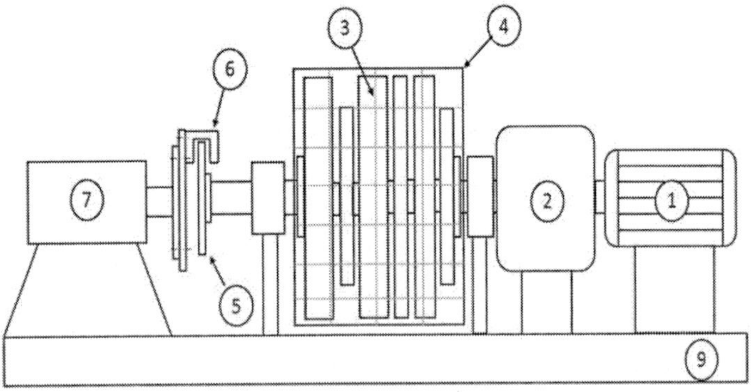

Figure 2. Single-end full-scale inertial dynamometer.

It is done in order to provide relatively average braking regimes in all wear tests (work done by brake application) under different brake interface temperatures. The brake interface temperature in the wear test 1 was maximum 100°C while in the wear test 2 it was 175°C. This range of change of the disc brake interface temperature has been considered as the most common and accordingly more relevant for wear testing of the friction materials. Gradual transition from transient to steady state wear conditions of the friction materials have been provided by the friction material burnishing (150 brake's applications). The number of brake's applications, for each wear test, has been chosen in order to measurable amount of wear be obtained in a short period (table 1). It is obvious that the ranges and distribution of the inputs data for training and testing have to be predefined. The ranges and distribution of data related to the brake operation regimes are defined by testing methodology (table 1). On the other hand, choice of the ranges and distribution of data related to composition and especially manufacturing parameters of the friction materials is difficult task. The ranges of ingredients used in the friction materials formulation are shown in table 2. Aramid, mineral, steel, and glass fibers have been included into the friction material formulations. Different friction modifiers have been used as lubricants in the friction material formulations such as graphite, friction dust, molybdenum disulphide. Moreover, typical abrasives in the friction material formulations have been used such as aluminium oxide, silica, and magnesium oxide. According to table 2, four different types of friction material (F1-F4) has been manufactured and tested for creation of a training data set.

Table 1 – Testing methodology

Test conditions	p (bar)	v (km/h)	T (°C)	Number of brake applications
Initial burnishing	40	90	<100	150
Wear test 1	40	90	≤100	100
Wear test 2	40	90	≤175	100

Table 2 – The ranges of types of ingredient used in friction material formulations (mass %)

Raw materials	F1-F4 (training data set)	F_T (test data set)
Phenolic resin	8-10	12
Fibers	26-40	35
Lubricants	12-19	17
Abrasives	8-11	2

On the other hand, type of friction material denoted as F_T has been tested using single-end full-scale inertia dynamometer in order to create a test data set. It is important to emphasize that the formulation of the friction material, which data has been used for testing the model abilities to predict how manufacturing parameters affecting wear of the brake friction material (see table 2). Furthermore, some input data used for testing have been out of the ranges used for the model development (phenolic resin, abrasives). From tables 2 and 3, it can be seen that five different types of friction material were produced as a disc pad assembly, mounted on the front brake (axle static weight of 730 kg) of a small passenger car and tested using the single-end full-scale inertial dynamometer. The disc pads with the friction surface area of 32.4 cm^2 and pad thickness of 16.8 mm were designed for the brake with an effective disc radius of 101 mm and floating calliper piston diameter of 48 mm. The formulation and manufacturing parameters for each type of friction material, as shown in tables 2 and 3, were completely different from one another.

Table 3 – The ranges of types of ingredient used in friction material formulations (mass %)

Manufacturing parameters	F1-F4 (training data set)	F_T (test data set)
Specific moulding pressure (MPa)	40-68.9	47
Moulding temperature (°C)	170	160
Moulding time (min)	10-11	7
Heat treatment temperature (°C)	200-260	230
Heat treatment time (h)	6-10	6

Results obtained during brake testing with friction materials F1-F4 were used for training of the artificial neural networks, while results with the friction material F_T were used for testing capabilities of the artificial neural networks. The manufacturing parameters, shown in table 3, have been randomly selected in the case of friction materials denoted as F1-F4. From table 3 it is evident that important manufacturing conditions such as moulding temperature and moulding time were out of the ranges used for the artificial neural network training. Furthermore, influence of the moulding pressure has been particularly investigated in the wide range of its change.

3. Modelling of Effects of Manufacturing Parameters

Computational methods based on the artificial neural networks can simulate the microstructure (neurons) of a biological nervous system. They can be trained to perform a particular function by adjusting the values of the connections (weights) between the elements. Each input to a neuron has a weight factor of the function that determines the strength of the interconnection and thus the contribution of that interconnection to the following neurons. As it was explained, in the field of tribology very complex and highly nonlinear phenomena are involved [22]. This is the reason why analytical models are difficult, even impossible to obtain. The analytical models of wear of brake friction materials were mostly made as functions load, sliding velocity, and sliding time [21,23]. Such defined models cannot be used for establishing a relationship between a formulation of brake friction material and particularly how manufacturing parameters affecting its wear properties. The improvement of friction material performance requires accurate modelling and prediction of the friction and wear processes versus friction material formulation, manufacturing, and testing conditions. That is why artificial neural networks are characterized as "computational models" with a particular ability to learn and generalize data from experimental data based on parallel processing [22].

It is clear that sufficient input/target pairs have to be stored in the training data set in order to artificial neural networks be learned about an input/output relationship. Input/output data have been obtained by formulation, manufacturing, and testing of five different types of friction material representing a data set that can be used for training and testing of the artificial neural network. The model of effects of manufacturing parameters on the friction material wear has been developed by taking into consideration 11 different input quantities (phenolic resin, fibers, abrasives, lubricants, moulding pressure, moulding time, moulding temperature, heat treatment time, heat treatment temperature, work done by brake, and brake interface temperature). The important input influence is related to work done by brake versus wear of the friction materials. Thus, the number of brake's applications has been prescribed per wear test (table 1), with each new brake application the set of input parameters have been changed. It is happened because the cumulative work done by brake applications are changing with each brake's application together with the initial brake interface temperature. Consequently, the linear wear i.e. wear volume of the friction material has also been changed with each brake's application. On the other hand, the linear wear of friction material becomes measurable after some number of brake's applications. The moment when liner wear of the friction material becomes measurable is unknown in advance. The linear wear of friction material was measured after all brake applications prescribed in the wear test (see table 1). Since synergistic effects of the input parameters on the wear of brake friction material are changed with each new brake application, 100 input values to the neural network is provided in the each wear test. Contrary, the linear wear of the friction material, as the output parameter, was measured after each wear test. It means that one value of the linear wear of friction material was provided at the end of wear test (see table 1). The two values of the output parameter i.e. the linear wear (wear volume) can be measured in two wear tests although 100 inputs have been changed per wear test. In order to provide enough wear results i.e. output values for neural networks training and to short the period of testing of the friction materials wear, the linear functional relationship has been assumed between work done per brake application and the number of braking. The same relationship was assumed between the number of brake's applications and wear of the friction material per brake application [23]. This means that the wear (w) achieved after one (first) brake application becomes ($2w$) after the second brake application, and $100w$ after 100 brake's applications in a wear test 1, for example. In order to develop the neural model able to predict and generalize the friction materials wear, the appropriate architecture of the neural network as well as the learning algorithm needs to be properly determined [22]. The best architecture to use depends on the kind of problem to be represented by the network. The trial and error method has been employed to find out the best artificial neural network characteristics for matching the particular input/output relationship. Based on MatLab 7.5 (R2007b), the two-layered artificial neural network architecture with five neurons in the first and two neurons in the second hidden layer, denoted as BR 11 $[5\text{-}1]_2$ 1, trained by Bayesian Regulation algorithm have been found as the best for functional approximation between 11 input quantities and one output value.

4. Influence of Manufacturing Conditions on Wear of Brake Friction Material

A commercial brake lining is manufactured by hot moulding of a mix under high pressure followed by heat treatments. The moulding is carried out for less than 15 min at a high pressure and temperature together with the heat treatment, which is performed for several hours with no pressure. The time, pressure and temperature during each step greatly affect physical and tribological properties of a brake lining [24]. That is why, it is very difficult to predict these influences on brake friction materials performance. Accordingly, requirements imposed to the model abilities to recognize and learn influences of the chosen manufacturing parameters are complex especially taking into consideration a formulation of brake friction material. The model based on neural computation, using artificial neural network, denoted as BR 11 $[5\text{-}1]_2$ 1, has been used for prediction how manufacturing parameters affecting wear volume of the new type of friction material (see table 2 and 3). Prediction of wear volume of the friction material F_T i.e. its specific wear rate, at brake interface temperature of 100°C (wear test 1), when work done by brake applications was 12827.047 KJ (after 100 brake's applications), is shown in Fig. 3. Furthermore, the same model has been used to predict a specific wear rate, at brake interface temperature of 175°C (wear test 2), when work done by brake applications was 12901.832 KJ. Comparison between the real and predicted values of wear volume i.e. specific wear rates related to the friction material F_T is shown in Fig. 3.

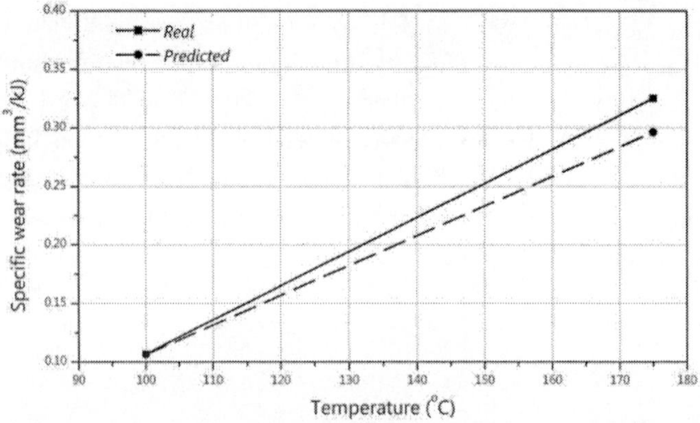

Figure 3 - Comparison between real and predicted specific wear rates (Reprinted with permission from SAE paper 2010-01-1679 © 2010 SAE International).

According to Fig. 3, it is evident that with increasing of the disc brake interface temperature, wear of the friction material F_T has been also increased. The model well predicted such behaviour of the friction material. From Fig. 3 it can be seen that increasing of the brake interface temperature for 75°C (wear test 1 and wear test 2) caused increasing of the friction material specific wear rate for approximately three times although almost the same cumulative work done by the disc brake in both wear tests. It means that wear of this type of friction material is sensitive versus brake interface temperature increasing. Increasing of the friction material wear could be explained by complex influence of formulation and manufacturing parameters used for the friction material production. In order to better identify how manufacturing conditions affecting the friction material wear properties, synergy of influences of the manufacturing conditions is shown in Fig. 4. From Fig. 4, the specific wear rates related to the friction materials used for training and testing at different brake interface temperatures can be seen. According to Fig. 4, the specific wear rates have been decreased with moulding pressure increasing in the range between 40 and 51MPa. Further increasing of moulding pressure of the friction materials, from 51 to 68.9MPa, for both brake interface temperatures, caused increasing of the specific wear rates (see Fig. 4). Obviously, change of moulding pressure substantially affecting wear properties of the brake friction material. It is shown in Fig. 4 that, for this type of brake friction material, increasing of moulding pressure from 40 to 51MPa, caused decreasing of wear. Regarding the character of change of manufacturing parameters shown in Fig. 4, it is evident that moulding pressure, moulding time, and heat treatment time have had the biggest influence of wear of the friction material. On the other hand, wear of the friction material denoted as F_T, cannot be analysed without considering its formulation parameters.

The neural model was tested against the data related to the friction material F_T, which manufacturing and formulation parameters have been changed versus those used for training. From table 2 it can be seen that this type of the friction materials manufactured with higher mass percentage of phenolic resin than other types of the friction material and lower percentage of abrasive ingredients. It could be expected that wear of this type of friction material become lower than other types of the friction material.

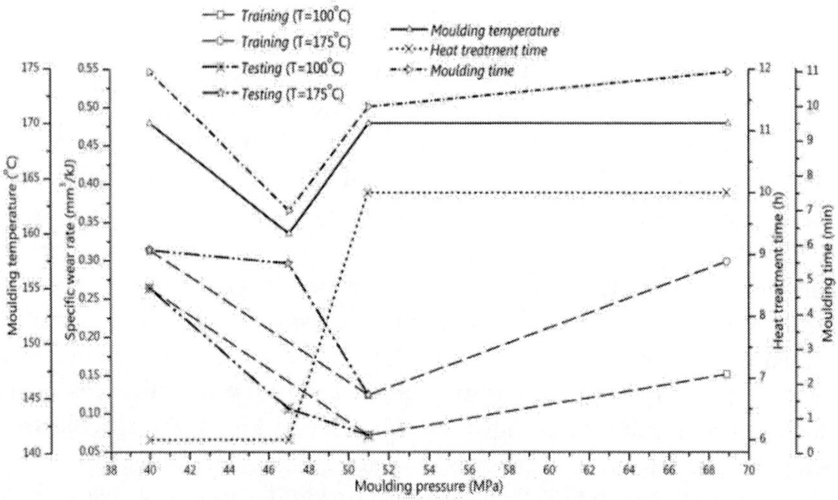

Figure 4 - Influence of the friction material manufacturing conditions on its wear at different brake interface temperatures (Reprinted with permission from SAE paper 2010-01-1679 © 2010 SAE International).

According to Fig. 3, the model well predicted decreasing of the specific wear rate at brake interface temperature of 100°C. The neural model predicted how the new set of manufacturing conditions related to moulding pressure of 47MPa, moulding temperature of 160°C, and moulding time of 7 min affecting the friction material wear. According to Fig. 4, it can be noted that further increasing of moulding pressure until 51MPa and simultaneously increasing of moulding time and temperature would cause further decreasing of the specific wear rate. On the other hand, at brake interface temperature of 175°C, the specific wear rate of the friction material F_T has been increased. This increasing of specific wear rate could be explained by lower values of moulding temperature (160°C) and time (7 min) as well as resin properties used for manufacturing of the friction material. The binder resin used for the friction materials producing has had the following properties: flow distance 32 mm, B-time 110s, hexa content 9%, and free phenol content 2%. It is important to know because flow distance is responsible for melt and flow behaviour of the resin during hot press moulding and degassing behaviour. It can be concluded that for higher brake interface temperature values, wear of the friction materials was primarily influenced by moulding pressure in synergy with moulding temperature and moulding time. From Fig. 4 it can be seen that decreasing of moulding pressure together with decreasing of moulding time and temperature caused increasing of specific wear rate of the friction material at brake interface temperature of 175°C. Furthermore, increasing of moulding temperature from 160°C to 170°C and moulding time from 7 min to 10 min caused decreasing of the specific wear rate at brake interface temperature of 175°C although the moulding pressure has being decreased. However, the wear volume of the friction materials cannot be analysed only versus manufacturing conditions but also the friction material formulations should be taken into consideration. It should be expected that this complex influence might be changed with change of brake interface temperatures. That is why it is very difficult to predict the nature of influence of this synergy.

If we look Figs. 4 and 5, in the range of moulding pressure from 40 to 47 MPa, when moulding temperature and time were decreased, specific wear rate of the friction material F_T has been decreased at brake interface temperature of 100°C while at brake interface temperature of 175°C it was increased. From Fig. 5, it can bee seen that in the same range of moulding pressure change, mass percentage of phenolic resin was increased while percentage of fibers decreased. The neural model well generalized such behaviour of the friction material F_T. It is interesting from fig. 4 that although the mass percentages of phenolic resin and fibers have been decreased, in the range of moulding pressure increasing between 47 and 51MPa, the wear volumes i.e. specific wear rates have been also decreased at both brake interface temperatures (see Fig. 5). Further increasing of moulding pressure for relatively constant values of manufacturing (moulding time, moulding temperature, heat treatment time) and formulation parameters caused increasing of the specific wear rates at both brake interface temperature. Based on that, it can be concluded that the specific wear rate of the friction materials strongly depends on moulding pressure for the specific friction material formulation. The influence of moulding temperature and time is especially evident at higher brake interface temperatures.

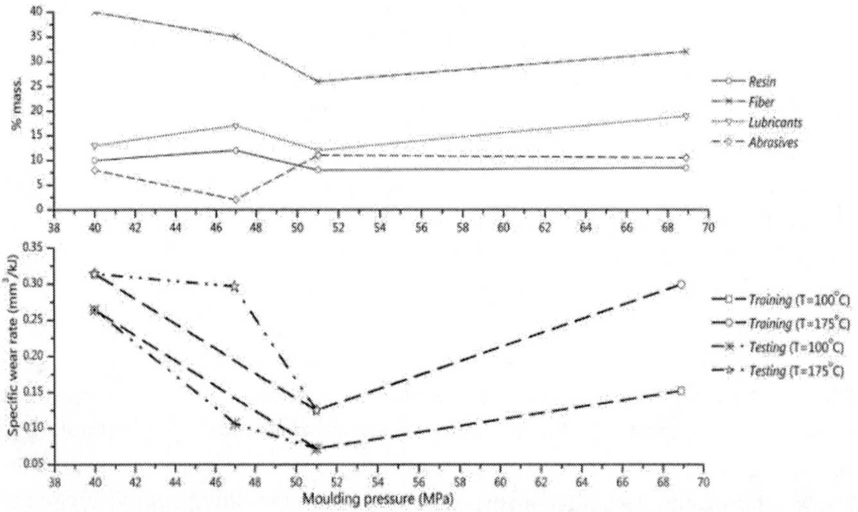

Figure 5 - Synergy of influence of manufacturing conditions of the friction material and its formulation on wear at different brake interface temperatures (Reprinted with permission from SAE paper 2010-01-1679 © 2010 SAE International).

Regarding Fig. 5, it can be seen that tuning of manufacturing and formulation parameters can substantially improve the friction material wear behaviour. If moulding pressure is increased from 47 to 51MPa, for this concentration of resin, abrasives, fibers, and lubricants, this moulding pressure together with increasing of moulding temperature on 170°C and moulding time on 10 min provided improving of wear properties of the brake friction material. It means that a specific wear rate, for brake interface temperature of 100°C, can be decreased on 0,07 mm^3/KJ. Moreover, for brake interface temperature of 175°C, this setting of manufacturing and formulation parameters can decreased a specific wear rate on 0,14 mm^3/KJ.

CONCLUSION

The influence of manufacturing conditions on wear of the friction materials is extremely important. Due to difficulties for use known analytical models of wear of brake friction materials, new an intelligent approach should be introduced. In this chapter, artificial intelligence was used to mine experimental data related to a formulation and manufacturing of the brake friction materials. Based on that, the neural model has been developed able to learn and generalized the character of influence of manufacturing parameters, such as moulding pressure, moulding time, moulding temperature, heat treatment time, and heat treatment temperature, on the brake friction material wear. Proper setting of manufacturing conditions of the friction material may cause substantial decreasing of wear volume of the friction material. This setting needs to be done taking into consideration of a formulation of the brake friction material. It was shown that the moulding pressure has had the most important influence on the friction material wear behaviour. Furthermore, it was shown how wear of some types of the friction material might be sensitive to change of the moulding pressure. Proper setting of moulding pressure may cause drastically decreasing of wear of the brake friction material. Additionally, fine tuning of moulding temperature and moulding time together with moulding pressure, may cause decreasing of wear. It was also shown that beside a moulding pressure, moulding temperature and time have significant influence on wear of brake friction material especially at higher brake interface temperatures. In this chapter was shown that increasing of moulding temperature from 160 to 170°C simultaneously with increasing of moulding time from 7 to 10 min and moulding pressure form 47 to 51MPa, caused significant improving of wear properties of the brake friction material. In order to adapt the manufacturing parameters of brake friction materials to a formulation of the brake friction material, the relationship between formulation, manufacturing, and tribo – behaviour of the brake friction materials should be modelled.

REFERENCES

[1] Rukiye E., Nurettin Y. An experimental study on the effects of manufacturing parameters on the tribological properties of brake lining materials, *Wear*, 2010, 268, 1524 – 1532.

[2] Kima Y.C., Cho M.H., Kim S.J., Jang H. The effect of phenolic resin, potassium titanate, and CNSL on the tribological properties of brake friction materials, *Wear*, 2008, 264, 204–210.

[3] Sasaki Y. Development philosophy of friction materials for automotive disc brakes, *Proc. JSAE 1995*, No. 9531679, 1995, 407-412.

[4] Jang H., Lee J. J., Kim S. J., Jung K.Y. The effect of solid lubricants on friction characteristics, *SAE*, No. 982235, 1998.

[5] Decuzzi P., Demelio G. The effect of material properties on the thermoelastic stability of sliding systems, *Wear,* 2002, 252, 311–321.

[6] Ostermeyer G.P. On the dynamics of the friction coefficient, *Wear,* 2003 254, 852–858.

[7] Eriksson M., Bergman F., Jacobson S. On the nature of tribological contact in automotive brakes, *Wear*, 2002, 252, 26–36.

[8] Aleksendrić D., Duboka Č. Prediction of automotive friction material characteristics using artificial neural networks-cold performance, *Wear,* 2006, 261, 269-282.

[9] Aleksendrić D., Duboka Č. Fade performance prediction of automotive friction materials by means of artificial neural networks, *Wear,* 2007, 262, 778-790.

[10] Schiffner K., Heftrich M., Brecht J. Modeling of Compaction Processes of Friction Material Mixes, 20th Annual Brake Qolloquium 2002, *SAE Paper 2002-01-2594*, Arizona, USA

[11] Archard J.F. Contact and rubbing of flat surfaces, *J. Appl. Phys*.,1953, 24, 981–988.

[12] Rhee S.K. Wear equation for polymers sliding against metal surfaces, *Wear* 1970, 16, 431–445.

[13] Ramalhoa A. Miranda J.C. The relationship between wear and dissipated energy in sliding systems, *Wear,* 2006, 260, 361–367.

[14] Abu Bakar A.R., Ouyang H. Wear prediction of friction material and brake squeal using the finite element method, *Wear,* 2008, 264, 1069–1076.

[15] Talib R.J., Muchtar A., Azhari C.H. Microstructural characteristics on the surface and subsurface of semimetallic automotive friction materials during braking process, *Journal of Materials Processing Technology,* 2003, 140, 694–699.

[16] Aleksendrić D., Duboka C. Fade performance prediction of automotive friction materials by means of artificial neural networks, *Wear*, 2007, 262, 778–790.

[17] Velten K., Reincke R., Friedrich K. Wear volume prediction with artificial neural network, *Tribology International,* 2000, 33, 731–736.

[18] Zhang Z., Friedrich K. Artificial neural networks applied to polymer composites: a review, *Composites Science and Technology,* 2003, 63, 2029–2044.

[19] Aleksendrić D. *Artificial neural networks in automotive brakes' friction material development,* PhD thesis, Faculty of Mechanical Engineering University of Belgrade, 2007.

[20] Aleksendrić D., Duboka C. Prediction of automotive friction material characteristics using artificial neural networks-cold performance, *Wear,* 2006, 261, 269–282.

[21] Aleksendrić D., Cirovic V. Effect of friction material manufacturing condition s on its wear, *28th Annual Brake Colloquium & Exhibtion 2010*, SAE Paper 2010-01-1679, October 10-13, Pheonix, Arizona, USA.

[22] Aleksendrić D., Duboka C. Automotive friction material development by neural computation, *BRAKING 2006,*York, United Kingdom.

[23] Aleksendrić D., Neural network prediction of brake friction materials wear, *Wear*, 2010, 268, 117-125.

[24] Seong J .K., Kwang S. K., Ho J. Optimization of manufacturing parameters for a brake lining using Taguchi method, *Journal of Materials Processing Technology,* 2003, 136, 202–208.

In: Manufacturing Engineering
Editors: Anthony B. Savarese
ISBN: 978-1-61209-987-3

Chapter 5

IMPACT OF SANCTIONS ON BUYER-SUPPLIER RELATIONSHIPS: A COMMENTARY OF A STUDY ABOUT THE LIBYAN OIL INDUSTRY

Jalal O. Tantoush, Fiona Lettice, and Hing Kai Chan
Norwich Business School, University of East Anglia, Norwich, Norfolk, UK

ABSTRACT

Buyer-supplier relationships have received increasing research attention in the last two decades. Empirical evidence shows that relationship quality can be examined by considering a number of relationship factors. Literature also shows that there are time-based features of relationship quality. Reviewed literature pinpoints a number of research gaps in this dyadic relationship context, one of which is the time-based feature context. Among others, empirical research gaps in this domain also include the geographical context.

Combining the geographical and time-based context research gaps, a pilot study explored buyer-supplier relationships between Libyan buyers and foreign suppliers over time (Tantoush *et al.*, 2009). The Libyan economy and in particular the Libyan oil industry was adversely affected by the imposition of sanctions by the US and the UN from 1986 to 2003. Libya's refining sector was hardest hit by UN Resolution 883 of 11 November 1993, banning Libya from importing refinery equipment. At the time, Libya was seeking a comprehensive upgrade to its entire refining system, with a particular aim of increasing output of gasoline and other light products (e.g. jet fuel).

The pilot study shapes the needs for further investigation in this area, and a proper research framework is needed in order to collect empirical data accordingly. This chapter aims to provide a commentary of the pilot study, and then guidelines for further research. More importantly, the reflections from this pilot study will be shared. The chapter will first briefly summarise the key findings of the pilot study. It will then present the implications and the recommendations for further study. After that, the benefits for conducting this pilot will be investigated.

INTRODUCTION

Buyer-Supplier Relationships

Dyadic relationships and buyer-supplier collaboration are gaining in theoretical and practical relevance. As Emberson and Storey (2006) note, one fairly typical definition of relationship marketing is that it denotes the 'establishing, developing and maintaining of successful relational exchanges' (Morgan and Hunt, 1994). Buyer-Supplier relationships are increasing in 'frequency and importance' (Smeltzer, 1997). As Fynes and Voss (2002) note, buyer-supplier relationships are issues that have attracted the attention of both academics and managers. According to Håkansson (1982), the area of buyer-supplier relationships has been the target of rigorous theory building and testing for many years, particularly within the industrial marketing literature, a fact which finds further support in Baglieri *et al* (2007).

The essential theme of such literature is the co-operative aspect of economic behaviour (Emberson and Storey, 2006). Substantial literature suggests that collaborative (or co-operative) relationships provide greater advantages than transactional relationships (e.g., Ganesan, 1994; Larson *et al.*, 2005; Kalwani and Narayandas, 1995; Daugherty *et al*, 2006). Cousins (2002) comments that a proper buyer-supplier relationship enables firms to be more flexible, adaptable and efficient. Undoubtedly, the attention given to it, by academics and managers, reflects its importance to organisational well-being.

Sanction on the Libyan Oil Industry

The UN and US sanctions had grave repercussions for the Libyan economy. Libya's refining sector was most severely hit as a result of the ban imposed on the importation of refinery equipment. Given these circumstances, buyer-supplier relationships in the Libyan oil industry suffered too. A review of the relevant literature on dyadic buyer-supplier relations and supply chain management highlights a paucity of research on the impact of sanctions on buyer-supplier relationships in this field (Tantoush *et al.*, 2009). While the issues surrounding dyadic buyer-supplier relationships have received increasing attention, little knowledge is available on the mediating influence of sanctions on such relationships.

In their study, Yahia and Saleh (2008) noted that most of the earlier literature has focused on the impact of economic sanctions mainly on the political consequences and the country's position in terms of supporting terrorism programmes. There has been little research on changes that could occur over time and under a sanctions regime in buyer-supplier relationships in a developing country. Sensitizing this critical context, this study aims to highlight the sanction and non-sanction era differentials of buyer-supplier relationships. Though scanty, sanctions literature flags the potential for differentials. Dani *et al* (2005) stress there is a need to highlight the possible time-based relationship differences given that, over time, different buyer-supplier relationship scenarios do emerge.

Connected to the above point, the Libyan economy, as highlighted earlier, is highly oil dependent. Insights about this relationship and sanction implications would offer key policy development substance. This chapter will provide a commentary note of a pilot study conducted by the authors earlier. The next section will briefly present the research method of the study, followed by a summary of the key results. Then, insights from the study will be suggested. More importantly, the benefits of the pilot study, in view of future research, will be revealed.

Research Method

The pilot study aimed primarily to: (1) pilot-test the conceptualised framework for this research direction; (2) identify some potential answers to the research questions for this study; and (3) identify the key issues that should be prioritised in the main study. Polit *et al.* (2001, p. 467) suggested that "a pilot study is a smaller-scale methodological trial intended to ensure that suggested methods and procedures will work in practice before being applied in a large, extensive investigation". Supporting this methodological logic, Baglieri *et al* (2007) suggest that a pilot study offers the opportunity to make adjustments and revisions before investing in, and incurring, the heavy burden associated with a large study. Thus, a pilot study is a trial before the main study.

A pilot study can refer to what is sometimes termed feasibility studies which are "small scale versions, done in preparation for the major study" (Polit *et al*, 2001, p. 468). According to Baker (1994, pp. 182-183), pilot research can also be the initial testing or 'trying out' of a particular research instrument. According to De Vaus (1993), one of the merits of conducting a pilot study is that it might give advance warning about where the main research project could fail, where research protocols may not be followed, or whether proposed methods or instruments are inappropriate or too complicated.

The Case Companies

Five companies were selected as the target case companies for the research. Key attributes of the companies are listed in Table 1. The reasons behind selecting the cases company was reported in Tantoush *et al.* (2009). In this chapter, the focus is to present the reflections from the pilot study and to propose possible future research directions in this area.

Table 1. Summary of the case companies

Company	Year of Establishment	Key Role in the Industry
National Oil Corporation (NOC)	1968	Oversee the marketing and production of oil and gas, developing exploration and production contracts and agreements.
Company A	1956	Involved in upstream activities, including oil exploration, drilling, production and shipping.
Company B	1960	Multinational oil and gas company.
Company C	2007	Leading gas distributor.
Company D	1981	Exploration, and production, manufacturing of oil and gas.

KEY RESULTS

Table 2. Summary of interview outcomes by theme (Tantoush *et al.*, 2009, p. 180)

Relationship Themes	NOC	Case Company A	Case Company B & Company C	Case Company D
Trust	During the sanction era the relationship was largely tense and devoid of mutual trust: suppliers were perceived as unreliable and untrustworthy. In the non-sanction era, mutual trust was restored.	The dyadic relationship reflected contractual trust (Cousins, 2002) during the non sanction period in contrast to the sanction period.	Mutual trust and reliability in the dyadic relationship during the non-sanction era in contrast to the sanction era.	Suppliers' relational trend showed nonchalance in relation to buyer's business interest in the sanction era (Lewin and Johnston, 1997)
Communication	While informal communication existed in the non-sanction era, completely rigid and formal communication prevailed in the sanction era. Similarly, communications were neither frequent nor timely (Fynes and Voss, 2002)	In contrast to the non-sanction era, communication from suppliers was perceived as neither timely nor meaningful (Anderson and Narus, 1990).	While the non-sanction era was characterized by smooth and timely communication, supplier dominance and censored communication was the trend during the sanction era.	Communication was neither frequent, timely (Morgan and Hunt, 1994) nor meaningful (Anderson and Narus, 1990) during the sanction era.
Cooperation	Goodwill and reciprocal support in the non-sanction era, and unfriendly and aggressive relational pattern during the sanction era.	Buyer-supplier relationships in the non-sanction era reflected 'mutual cooperation' while the supplier dominated in the sanction era.	The relationship trend in the non-sanction era was 'interdependent' and 'supplier dominant' (Dani *et al*, 2005) in the sanction era.	In the sanction era, the supplier showed unwillingness to resolve problems quickly and aid the buyer's relational interest.
Commitment	The dyadic relationship in the sanction era was characterized by attitudinal commitment that sacrificed the buyer's goal (Gundlach *et al*, 1995), in contrast to the non-sanction era.	In contrast to the non-sanction era, the suppliers showed low commitment to the relationship.	The suppliers showed significantly less loyalty and involvement to help the buyer's interest (Gundlach *et al*, 1995) during the sanction era	In an attitudinal commitment context, suppliers' behavioural trend depicted a lack of intent to commit time and resources to the buyer's plans (O'Reilly and Chatman, 1986).

Pilot studies can be based on quantitative and/or qualitative methods. The pilot study explained here is qualitative and case study based, and data was collected through 5 in-depth, semi-structured interviews (lasting between 90 and 120 minutes) and examination of documents collected during case studies. A summary of the key outcomes of the interviews is provided in Table 2. Again, detailed discussions of the themes in Table 2 are reported in Tantoush *et al.* (2009) and thus they are not reproduced in this chapter. Nevertheless, we focus our discussions on the sanction-based analysis.

Sanction-Based Findings

Overall, interview results and secondary data (e.g., company memos and meeting records, where available) from all respondent companies showed that there were sanction-based influences on their relationships. In other words, there were significant differences in the relationship between the buyers and suppliers during the sanction and non-sanction periods. Not surprisingly, the Libyan companies and their foreign suppliers were closer during the non-sanction period. What this study shows is what factors affect the relationship and how the relationship changed during the sanctions period.

From Company B and Company C's evidence, unlike in the non-sanction period, the suppliers (dealers in spare parts) required end user certificates. These had to be sent to an inspector to ensure the parts were being used specifically for the oil industry and not for a nuclear programme, during the sanction period. According to the respondents of Company B and Company C, some suppliers showed some support for Libyan companies during the sanction period because these companies had European partners, such as Italy.

Some suppliers, who also had been dealing with Libya before the sanction period changed their procedures from letters of credit to cash up front. In very rare cases, new supplier relationships, which were developed during the sanctions period, worked well in the buying and supplying of required orders. This is supported by Company B, Company C, and Company D.

While noting that there were several other relationship quality differences between the sanction and non-sanction periods, one respondent preferred to stress spare parts procurement sanction-based issues. Concerning buyer-supplier problems during the sanction period, the spare parts were more expensive than during the non-sanction period, he noted. Company A had a Libyan-American partnership, but the US froze their shares from 1986 until the sanctions were finally lifted in 2003. Further interview results showed that Company A spent enormous amounts of money on tools and spare parts, which were not all used during the sanctions period. However, despite the huge costs involved, they could not stop the production line operations due to contract commitments. At the time sanctions were lifted, there was still excess stock and a lot of money wasted.

In addition, Company D found that when the US left the company, some of their suppliers cut off all relationships. They had to buy treble the number of important spare parts to keep oil operations going in order to honour the many contracts they had signed. This resulted in greater expenses to keep their inventories stoked up. If they did not supply the customer with oil then they would pay a penalty for the delay. They bought some spare parts through the NOC office in London or Germany. The company faced some obstacles to finishing projects on time because of supplier issues. However, orders started with a partner before the sanctions period, were finished on time because of trust, communications and cooperation with the supplier.

Interviews with the NOC also revealed sanction-based differences. According to the NOC Head of Planning Department respondent, they stocked up their spare parts during the sanction period either through London or Germany. Some parts were also obtained from China at a lower price. The Chinese market, during the sanctions period opened new doors for business and a new market for the Libyan oil industry. While some European suppliers supported the Libyan companies during the sanctions, their prices were very high compared to the Chinese suppliers. As the NOC respondent also reported, after the sanctions were lifted there was huge competition between suppliers to win Libyan oil contracts for providing tools and spare parts.

INSIGHTS FROM THE STUDY

Theoretically, the pilot study offers a number of insights which serve as a guide for the future study. Two main groups of theoretical insights were captured, namely sanction-based, and relationship quality theme-based. These insights are summarised below:

The Sanction and Non-Sanction Differentials

There were significant differentials in the relationship quality in the buyer-supplier relationships between the sanction and non-sanction periods. Thus, the imposition of sanctions had serious impacts on the relationship between the buyer companies (buyers) and the foreign companies (suppliers). Evidence from the pilot study on the sanction context, also showed the need to deeply explore the sanction impact on the relationships, to pinpoint relevant theoretical backgrounds.

The pilot study supported existing literature that suggests trust, cooperation, communication, commitment, satisfaction and pricing as core themes for examining relationship quality. The relationship quality themes are summarised next.

Trust

For the trust theme, all pilot study case study companies identify trust as a core feature of relationship quality. Also, they note that trust levels will differ between the sanction and non-sanction periods. Putting interview results and company documentary evidence together, the characteristics of trust would include reliability, frankness, and showing concern for the other's welfare.

Cooperation (Collaboration)

Also, interview results showed that in the sanction period, cooperative behaviour was lacking in the explored relationship, unlike in the non-sanction period. From combined evidence, the characteristic features of this theme to be transported to the main study includes showing willingness to cooperate, caring for the other party's profitability, and showing willingness to allow for bargaining.

Communication

The pilot study showed that the partners in a buyer-supplier relationship must engage in effective communication, as this is very important to understanding each other and agreeing their targets. Overall, combining evidence from interviews and written documentation, explored companies support the need to emphasise regularity and timeliness of communication, as well as informal and formal types of communication, in their relationship with the respective suppliers.

Commitment

One other theoretical take from the pilot study that is transported to the main study is the evidence that commitment is a core factor of buyer-supplier relationships. From the empirical evidence, this factor needs to be further explored in the main study.

From the pilot study results, additional to this importance to explore commitment in buyer-supplier relationships, three core commitment features were identified for use as guide in the main study. These include dedication to the relationship, display of patience in behaviour towards the other party, and investment in the relationship.

Satisfaction

Another insight from the pilot study is that satisfaction is an important relationship factor and needs to be explored. All explored companies supported this importance, and identify economic and social interaction elements of buyer-supplier relationships.

Pricing

Pricing is also important in the examination of buyer-supplier relationships. In addition to pinpointing the core importance of this factor, respondents of explored companies also underline key pricing behaviours during sanction periods. These backgrounds are transported to the main study.

In conclusion, this pilot study shapes the needs to address the following questions:

What are the core relationship quality factors in the buyer-supplier relationships between Libyan oil companies (Buyers) and their European and American partners (Suppliers)?

What are the differentials in the relationships between the sanction and non-sanction periods? What are the sanction implications for the relationships?

BENEFITS OBTAINED FROM THE PILOT STUDY

Research Methodology-Based Benefits

A summary of research methodology-based benefits, part of which is also supported by previous studies, of the pilot study in this research is shown below:

It enabled the authors to test the appropriateness of the topic guide, sentence structures and probing techniques (De Vaus, 1993; Frankland and Bloor, 1999).

It helped in designing the main study research protocol (Polit *et al*, 2001; Tashakkori and Teddlie 1998).

Assessing whether the research protocol was realistic and workable (Polit *et al*, 2001). It was on the basis of the pilot studies that the authors realised that conducting focus group discussions was imperative for the successful completion of this research.

Establishing whether the sampling frame and technique are effective.

Assessing the likely success of proposed recruitment approaches and recognising that interviewees needed to be reminded at least twice before the main interviews (Tashakkori and Teddlie, 1998).

Identifying logistical problems which might occur.

Estimating variability in outcomes, which helped to better estimate the sample size for the main study.

Collecting preliminary data. Also, this pilot study has enabled the researchers to better understand the data collection process for the main study and has identified potential areas that need more effort to improve the research process.

Help in determining what resources were needed for the main study.

Assessing the proposed data analysis techniques to uncover potential problems.

It assisted with the further development of the research question and research plan for the main study. It has also helped to define and refine the interview questions and provide a stronger focus for the main study.

It also enabled the estimation of expected interview duration.

Theoretical Focus Benefits

It is worth mentioning that the benefits gained from the pilot case studies were not limited to testing interview skills and gaining interviewing experience. The pilot interviews provided valuable insights into understanding issues of the impact of sanctions on the buyer-supplier relationships within the Libyan oil industry.

Insights from the pilot study contributed considerably to shaping the theoretical focus of the main study interviews. Not only did the insights help in the main study investigation of the sanction and non sanction issues, but also on the themes and features of relationship quality.

CONCLUSION

Conducting the pilot study was very helpful for the main study. Overall, doing the pilot study offered two main streams of benefits as mentioned in the preceding section. However, the authors would like to stress that although pilot study findings might offer some indication of some issues concerning the main research, they cannot guarantee the scale of success of the main study. This is because they do not have a generalised foundation and are nearly always based on small numbers. Furthermore, other problems or hitches may not become obvious until the larger scale study is conducted. Nevertheless, where an accomplished and validated technique is being used and the pilot study is determining other methodological aspects, it could be argued that such data may be of value, which is the intention of presenting this chapter.

REFERENCES

Anderson, J., & Narus, A. (1990). A model of distribution firm and manufacturer firm working partnerships. *Journal of Marketing*, 54, 42-58.

Baglieri, E., Secchi, R., & Croom, S. (2007). Exploring the impact of a supplier portal on the buyer–supplier relationship. The case of Ferrari auto. *Journal of Industrial Marketing Management,* 36, 1010–1017.

Baker, T. L., Simpson, P. M., & Siguaw, J. A. (1999). The impact of suppliers' perceptions of reseller market orientation on key relationship constructs. *Journal of the Academy of Marketing Science*, 27(1), 50-57.

Cousines, P. D. (2002). A conceptual model for managing long-term inter-organisational relationships. *European Journal of Purchasing & Supply Management,* 8(2), 71-82.

Daugherty, P. J., Richey, R. G., Roath, A. S., Min, S., Chen, H., Arndt, A. D., & Genchev, S. E. (2006). Is collaboration paying off for firms? *Business Horizons,* 49, 61–70.

Dani, S., Burns, N. D., & Backhouse, C. J. (2005). Buyer–supplier behaviour in electronic reverse auctions: a relationships perspective. *International Journal of Services and Operations Management,* 1(1), 22-34.

De Vaus, D.A. (1993). *Qualitative Research in Social Research* (3rd edition). London: UCL Press.

Larson, P. D., Carr, P., & Dhariwal, K. S. (2005). SCM involving small versus large suppliers: Relational exchange and electronic communication media. *Journal of Supply Chain Management*, 41(1), 18-29.

Emberson, C., & Storey, J. (2006). Buyer–supplier collaborative relationships: beyond the normative accounts. *Journal of Purchasing and Supply Management,* 12, 236-245.

Frankland, J., & Bloor, M. (1999). Some issues arising in the systematic analysis of focus group material. In: R. Barbour and J. Kitzinger (Eds),

Developing Focus Group Research: Politics, Theory & Practice. London: Sage

Fynes, B., & Voss, C. (2002). The moderating effect of buyer-supplier relationships on quality practices and performance. *International Journal of Operations and Production Management*, 22(6), 589-613.

Ganesan, S. (1994). Determinants of long-term orientation in buyer/seller relationships. *Journal of Marketing*, 58, 1-19.

Gundlach, G. T., Achrol, R. S., & Mentzer, J. T. (1995). The structure of commitment in exchange. *Journal of Marketing*, 59, 78-92.

Håkansson, H. (Eds.) (1982). *International Marketing and Purchasing of Industrial Goods.* Chichester: Wiley.

Kalwani, M., & Narayandas, N. (1995). Long-term manufacturing-supplier relationships: do they pay-off for supplier firms? *Journal of Marketing*, 59(1), 1–16.

Lewin, J. E., & Johnston W. J. (1997). Relationship marketing theory in practice: a case study. *Journal of Business Research*, 39, 23– 31.

Morgan, R. M., & Hunt, S. D. (1994). The Commitment-Trust theory of Relationship marketing. *Journal of Marketing* 58, 20-38.

O'Reilly, C. A., & Chatman, J. (1986). Organizational commitment and psychological attachment: The effects of compliance, identification and internalization on prosocial behaviour. *Journal of Applied Psychology*, 71, 492-499.

Polit, D. F., Beck, C. T., & Hungler, B. P. (2001). *Essentials of Nursing Research: Methods, Appraisal and Utilization (5th edition).* Philadelphia: Lippincott Williams & Wilkins.

Smeltzer, L. (1997). The meaning and origin of trust in buyer-supplier relationships. *International Journal of Purchasing and Materials Management,* 33(1), 40-48.

Tantoush, J. O., Lettice, F., & Chan, H. K. (2009). The impact of sanctions on buyer-supplier relationship within the Libyan oil industry. *International Journal of Energy Sector Management* 3(2), 171-186.

Tashakkori, A., & Teddlie, C. (1998). Applied Methodology: Combining Qualitative & Quantitative Approaches. London: Sage.

Yahia, A. F., & Saleh, A. S. (2008). Buyer-seller relationships. *American Journal of Applied Sciences*, 5(12), 1713-1719.

In: Manufacturing Engineering
Editors: Anthony B. Savarese
ISBN: 978-1-61209-987-3

Chapter 6

PROCESSING AND CHARACTERIZATION OF THE ALLOY NB46WT%TA POWDER FOR USE IN ELECTROLYTIC CAPACITORS

J.N.F. Holanda, U.U. Gomes and D.G. Pinatti
[1]Northern Fluminense State University, Laboratory of Advanced Materials/Group of Ceramic Materials, Campos dos Goytacazes, RJ, Brazil
[2]Federal University of Rio Grande do Norte, Department of Theoretical and Experimental Physics, Natal, RN, Brazil
[3]University of São Paulo, Engineering College of Lorena, Lorena, SP, Brazil

ABSTRACT

The electrolytic capacitor is an electronic device of great technological importance because it is widely used in electrical circuits. Electrolytic capacitors are used in electronic equipments of defense, information technology, industrial communication, tools for office, automation, robotics, etc. The preference for electrolytic capacitor is related to its high specific capacitance (high value of capacitance in small volume). Tantalum is the main material used for manufacturing of electrolytic capacitors of high performance. Only its high costs restrict a wider application. This fact has stimulated scientific research in order to establish a lower cost substitute material. This chapter focuses on the production and characterization of the alloy Nb46wt.%Ta powder. The substitution of pure tantalum by the alloy Nb46wt.%Ta results in economical benefits. The reasons are: i) the substitution of tantalum by niobium by 46 wt.% results in a cheaper material for electrolytic capacitors; and ii) powder production of the alloy Nb46wt.%Ta using aluminothermic step and subsequent comminution operation is cheaper.

Keywords: alloy Nb46wt.%Ta, powder, electrolytic capacitor, sintering, electrical properties.

1. INTRODUCTION

The capacitor is considered one of the most important electronic components employed in the microelectronics industry. The capacitor is widely used in electric circuits, which are very important to modern society. In fact, a circuit consists of electrical elements linked together in a closed path so that an electric current may continuously flow [1].

In its simplest form, the capacitor is a two-terminal element that is a model of a device consisting of two conducting plates separated by a dielectric material [2], whose primary purpose is to introduce capacitance into an electric circuit. The capacitor has stored energy by virtue of the separation of charges between the capacitor plates. These charges have an electrical force acting on them. Electrical charges are stored on the plates.

The capacitance value is proportional to the surface area of the plates and dielectric constant, and is inversely proportional to the thickness of the dielectric material. For the configuration of a capacitor of parallel-plates the capacitance (C) can be computed from the relationship [2]:

$$C = \varepsilon_0 . \varepsilon . A / X \qquad (1)$$

in which ε_o is the dielectric constant of a vacuum (8.85 x 10^{-12} F/m), ε is the dielectric constant of the material, A is the surface area of the plates, and X is the thickness of the dielectric material between the plates. According with equation (1) to obtain greater capacitance, a very thin dielectric material, high dielectric constant, and plate of large surface are required.

The ability to store energy in a capacitor is measured by the value of capacitance (C). In the SI the unit of capacitance is the Farad (F). Thus, 1F is defined as 1 Coulomb/Volt. From a practical standpoint, the most convenient units are sub-multiplies of Farad (1 μF = 10^{-6} F and 1 pF = 10^{-12} F). Capacitors come in a wide range of values between 1 pF to 0.22 F.

The capacitors use various dielectric materials and are built in several forms [3]. Some common electric capacitors use impregnated paper for a dielectric material, whereas others use ceramic materials and plastics. The values of the dielectric constant for several dielectric materials are given in Table 1 [2,4,5].

Table 1. Dielectric constant for some dielectric materials [2,4,5]

Material	Range
Vacuum	1.000000
Air	1.000054
Water	78
Paper	3.5
Mica	5.4
Porcelain	6.0
Soda-lime glass	6.9
Fused silica	4.0
Nylon 6.6	4.0
Polystyrene	2.6
Polyethylene	2.3
TiO_2	100
Ta_2O_5	27
Nb_2O_5	41
Al_2O_5	10
$BaTiO_3$	1,000 – 5,000
$Ba(TiZr)O_3$	~ 10,000
$Pb(Mg_{1/3}Nb_{2/3})O_3$	~ 18,000

The electrolytic capacitor is of great technological importance due to its high specific capacitance, i.e., high capacitance value in small volume. This type of capacitor is the one with the highest capacitance per unit volume at low cost. The materials commonly used in the manufacture of electrolytic capacitors are tantalum and aluminum [6,7].

The tantalum is the preferred material for application in electrolytic capacitors of high performance [8]. This is related to the following factors: i) high specific capacitance due to the finesses with which tantalum powder can be produced; ii) a very stable oxide film that can perform as a dielectric; and iii) a high relative dielectric constant of about 27 for Ta_2O_5. These factors lead to the main application of tantalum as an anode material of electrolytic capacitors, which are used to provide relatively large capacitance values or relatively small physical size. The tantalum capacitors are widely used in military systems, space programmers, non-entertainment automotive applications, precision electronics, information technology, hybrid microelectronics, RC networks, bypass or decoupling digital circuits, among others [7, 9-11]. In addition, the tantalum capacitors have a very high reliability in a wide range of environments, an extremely important factor.

The tantalum electrolytic capacitor of high performance is produced using tantalum powder, with certain quality requirements described by its physical, chemical, mechanical and electrical characteristics. There are two types of tantalum capacitor powders: 1) sodium-reduced powders [12, 13]; and ii) powders produced from electron beam fused (EB) tantalum metal [14]. Sodium-reduced powders present high specific charge, and are generally used for applications in the lower voltage range up to 35 V. The EB powders present low specific charge, and are used in the high voltage range between 35 – 120 V.

The tantalum capacitor has over time been one of the most important components used in electrical circuits [15]. Only its high cost restricts a wider application. For this reason, the tantalum capacitors in some applications are being substituted by ceramic and aluminum capacitors. This situation has stimulated technological research with the objective of finding a substitute material of lower cost.

Several materials such as niobium, titanium, hafnium, zirconium and alloy of the Ti-Al system have been investigated as a substitute material for the tantalum [16-19]. However, these materials present performance that are much lower than those found in tantalum. Alloys of the Nb-Ta system, obtained by the simultaneous aluminothermic reduction of Nb_2O_5 and Ta_2O_5, have been investigated as alternative to tantalum [15,20,21]. These alloys are attractive for application in electrolytic capacitors because of the following factors: i) lower cost of aluminothermic alloys; ii) possibility of production of porous sintered anodes with high specific surface area; iii) the relative dielectric constant of the Nb-Ta oxide films is relatively high; and iv) the Nb-Ta alloys have about half the density of tantalum. In particular, the alloy Nb46wt.%Ta presented promising results for application in electrolytic capacitors [20].

This chapter focuses on the production and characterization of alloy Nb46wt.%Ta powder for use in electrolytic capacitors. Emphasis is put on the sintering behavior and electrical properties of the sintered anodes as a function of the sintering conditions.

2. Experimental Procedures

The alloy Nb46wt.%Ta used in this work was produced by the simultaneous aluminothermic reduction of the oxides Nb_2O_5 and Ta_2O_5 in the adequate proportions [20], and was supplied by the FAENQUIL-DEMAR (Lorena-SP, Brazil). The different process steps involved in the powder production and characterization are shown schematically in Fig. 1.

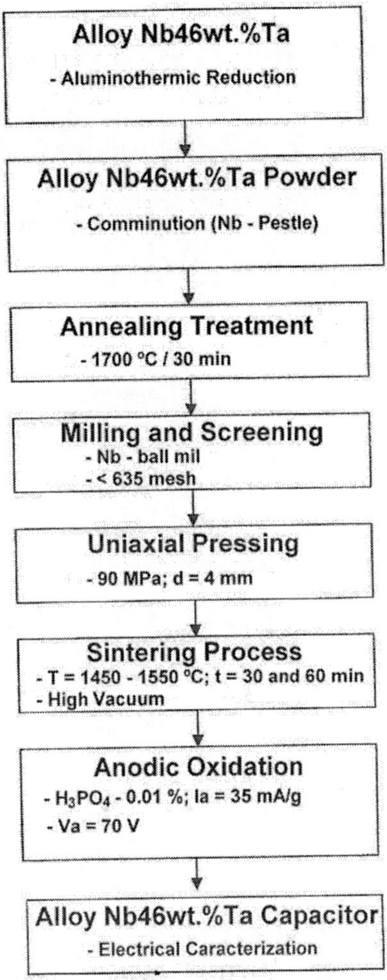

Figure 1. Process flow diagram of the alloy Nb46wt.%Ta capacitors.

The alloy Nb46wt.%Ta powder was produced by mechanical comminution. The alloy was ground to powder in a pestle of Nb and subsequent wet ground in a planetary ball mill, using a milling time of 6 h. The cup and milling balls were made of niobium.

The produced powder was submitted the annealing treatment at 1700 ºC/30 min. This treatment has as goal to purify the alloy Nb46wt.%Ta powder with respect of aluminum. In addition, it is expected that the annealing treatment process create a superficial porosity at the powder particles. The powder was than classified by the sieving down to < 635 mesh (< 20 μm). The chemical composition of the alloy Nb46wt.%Ta is given in Table 2. As observed in Fig. 2, the alloy Nb46wt.%Ta powder consists of angular-shaped particles. This morphology results from cleavage of brittle alloy Nb46wt.%Ta particles during comminution.

Table 2. Chemical composition of the alloy Nb46wt.%Ta powder (ppm)

Elements	Range
N	1074
O	6016
C	468
Fe	1250
Si	2000
Ti	990
Mo	149
Al (%)	5.48

Figure 2. Morphology of the alloy Nb46wt.%Ta powder particles.

Samples for sintering were prepared by single action uniaxial pressing at 90 MPa in a cylindrical 4 mm diameter steel die. All samples were fitted with a 0.5 mm niobium wire to serve as an electrical contact for anodic oxidation and subsequent electrical measurements.

The compacted samples were sintered in a resistance heated furnace under high vacuum (< 10^{-5} Torr.), using niobium sheets as a heating element. The applied sintering temperatures were 1450 °C, 1500 °C and 1550 °C for 30 and 60 min. Before and after sintering the outer dimensions of the samples were measured.

Anodic films of well-defined thickness were grown on the sintered samples by anodic oxidation at 70 V by applying a current density of 35 mA/g. An aqueous solution of H_3PO_4 (0.01 wt.%) at 90 °C was used as the electrolyte. The formed anodic film corresponded to the dielectric medium, which was between the plates of a capacitor.

Capacitance, dissipation factor and equivalent series resistance measurements were determined at 120 Hz, 1 kHz and 10 kHz by using an HP-LCR 4262A equipment. An aqueous solution of H_3PO_4 (10 wt.%) at 23 °C was used as the electrolyte. The bias voltage used was of 2.2 V. The specific surface area and specific surface area reduction of the sintered samples were determined through electrical capacitance measurements at 120 Hz.

Leakage current measurements were determined by using an electrometer (model 614, Keitley). An aqueous solution of H_3PO_4 (10 wt. %) at 23 °C was used as the electrolyte. The bias voltage used was of 2 V during 2 min.

Breakdown voltage measurements were determined by using a stabilized voltage source. An aqueous solution of H_3PO_4 (10 wt. %) at 90 °C was used as the electrolyte.

3. RESULTS AND DISCUSSION

3.1. Sintering Behavior of the Alloy Nb46wt.%Ta Powder

The linear shrinkage ($\Delta L/Lo$) for all the samples of alloy Nb46wt.%Ta powder as a function of the sintering conditions is shown in Fig. 3. It was observed that the samples presented low shrinkage (0.4 – 0.6 %) at 1450 °C, mainly for the samples sintered at 30 min. This indicates that probably the surface diffusion is the dominant sintering mechanism. At 1550 °C the samples presented higher shrinkage (~ 1.6 %). This result suggests that in this temperature range there is a change in the sintering mechanism.

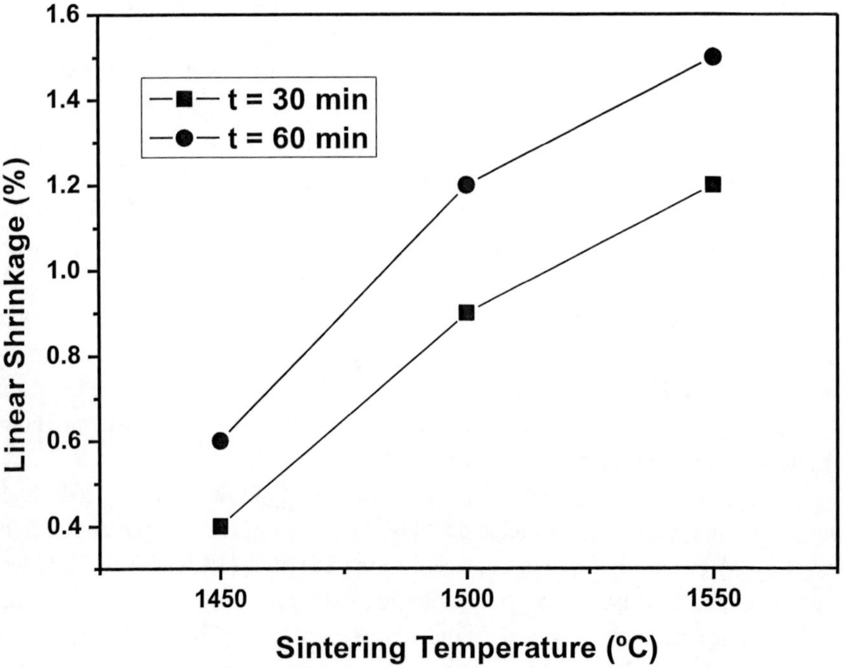

Figure 3. Linear shrinkage of the samples sintered at different temperatures.

The specific surface area of the samples as a function of the sintering temperature and time is shown in Fig. 4. As can be observed, the samples sintered at 1450 °C for 30 min presented low value of surface area. This behavior can be related mainly to the following factors: i) in this sintering condition the purification of the anodic surface didn't occur yet; ii) the neck growth wasn't sufficient to give adequate properties to the anodic oxide film formed by the anodizing process; and iii) the formation of the conductor electrically ways in the sintered samples were inhibits, in which in all powder particles were not anodized. At 1450 °C for 60 min, however, the highest value of specific surface area was obtained. This result is related to the volatization of the surface impurities, and mainly the creation of superficial porosity at particles of the alloy Nb46wt.%Ta powder obtained via aluminothermic reduction process. In addition, this sintering condition isn't sufficient for provoke total sintering of the samples. As sintering proceeds (1500 °C and 1550 °C) the specific surface area of the samples becomes smaller.

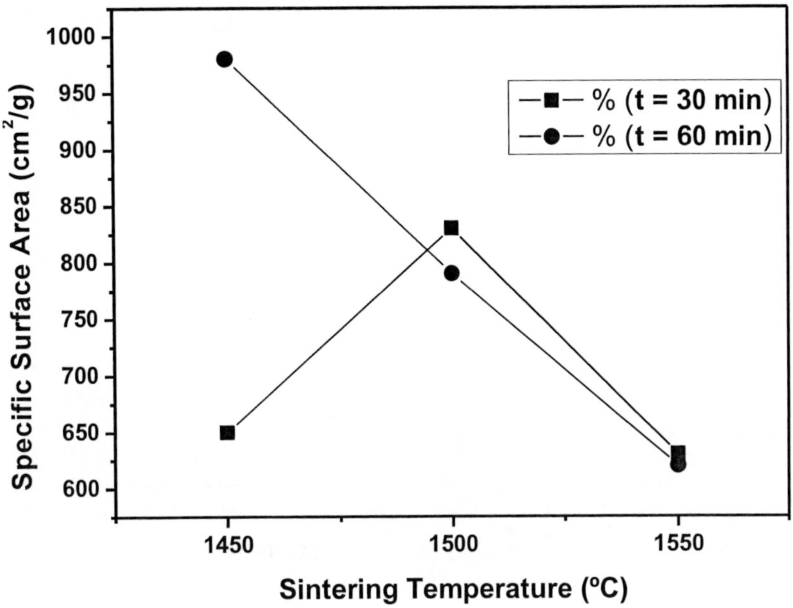

Figure 4. Specific surface area of the samples sintered at different temperatures.

The specific surface area reduction (ΔS/So) of the samples as a function of the sintering temperature and time is shown in Fig. 5. The samples should maintain maximum porosity after sintering, once this fact determines the capacitance [22]. Thus, the reduction of the specific surface area of the samples should be kept as low as possible. It was observed that the samples sintered at 1450 °C presented lower value of ΔS/So. Above 1450 °C, however, high specific surface area reduction occurred, mainly at 1550 °C (ΔS/So = 42 %). This can be explained due to the alloy Nb46wt.%Ta powder used presents a small particle size (< 20 μm).

Fig. 6 shows the plot of the linear shrinkage (ΔL/Lo) versus specific surface area reduction (ΔS/So) for the samples sintered at 60 min. The correlation between ΔL/Lo and ΔS/So is well established, in which the correlation coefficient presented value of $R \rightarrow 1$. This result apparently indicates that between 1450 °C and 1550 °C a change in the dominant sintering mechanism is not occurring.

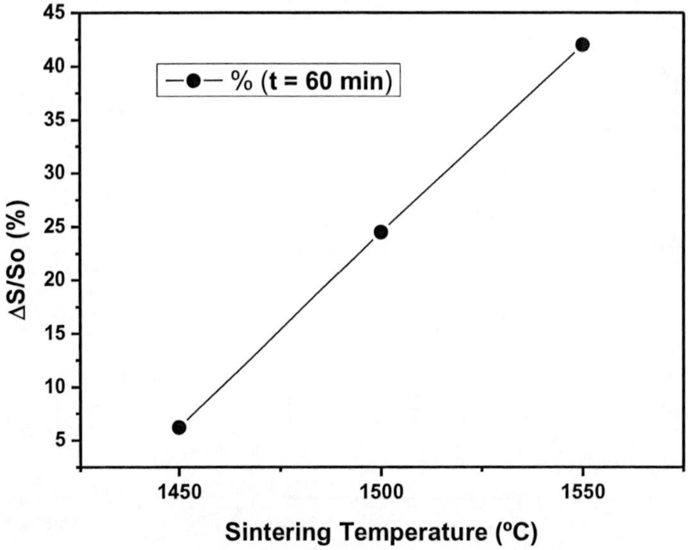

Figure 5. Specific surface area reduction of the samples sintered at different temperatures.

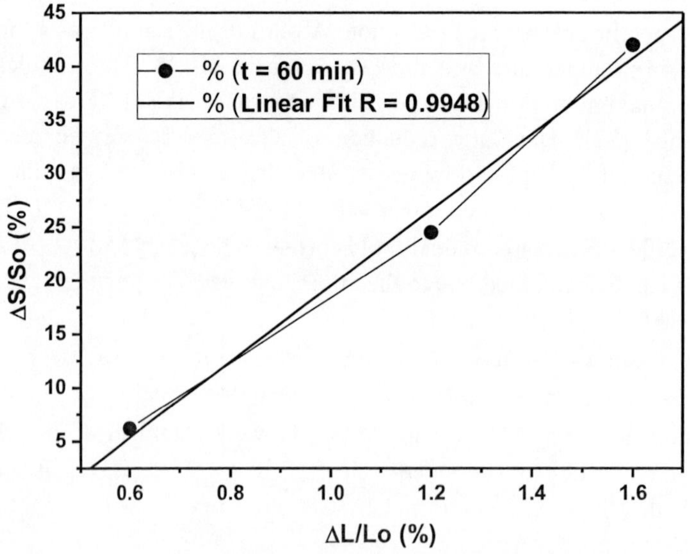

Figure 6. ΔL/Lo x ΔS/So for the samples sintered during 60 min.

From the plot log (ΔS/So) versus log (t) for the samples sintered at 1500 °C, the solid sintering mechanism (γ) was calculated [23]. It was obtained a value of γ = 3.7. This result indicates that for the studied conditions the sintering process of the alloy Nb46wt.%Ta powder is surface diffusion controlled.

3.2. Electrical Characterization of the Alloy Nb46wt.%Ta Capacitor

The data in Table 3 show the specific capacitance for the alloy Nb46wt.%Ta capacitor obtained under different sintering conditions. All the capacitance measurements were obtained at different frequencies. It was found that the specific capacitance is high relatively, and varies strongly with the sintering conditions. In general, when the sintering temperature and time are increased, the surface, and thus the capacitance, decreases. On the other hand, it was used alloy Nb46wt.%Ta powder with small particle size (< 20 μm) that leads to higher capacitance. In fact, the capacitance increases with the decreasing of the particle size according to [24]:

$$C = 4.90 \times 10^7 \, S.h / d.Va \qquad (2)$$

in which C is the capacitance, S is the surface area of the anode, h is the length of the anode, d is the particle size, and Va is the anodizing voltage.

Table 3. Results of specific capacitance for the alloy Nb46wt.%Ta capacitor sintered at different sintering temperatures and times

Sintering		Specific Capacitance (μF/g)		
T (°C)	T (min)	120 Hz	1 kHz	10 kHz
1450	30	161	88	34
1450	60	236	75	23
1500	30	200	77	10
1500	60	163	70	7
1550	30	160	88	34
1550	60	152	92	40

At 1450 °C for 60 min, the highest value of specific capacitance (236 μF/g) was obtained. This can be explained, because under this condition low densification (Fig. 3) and highest specific surface area (Fig. 4) were observed. In this case only the initial link between powder particles with a brittle neck structure was formed, resulting in a porous sintered anode. As expected, the values of capacitance decrease with the increase of the frequency for all sintering conditions.

Table 4 shows the values of the dissipation factor (Tg δ) for the alloy Nb46wt.%Ta capacitor obtained under different sintering conditions. The measurements of the dissipation factor Tg δ give information on the quality of the dielectric film grown on the alloy Nb46wt.%Ta. It can be observed that as sintering proceeds (higher temperature or longer time) the value of Tg δ decreased. This behavior could be related to the purification of the anodic surface by the impurities volatization. In addition, at higher temperatures the sintered anodes are less porous. At 1550 ºC for 60 min the lowest dissipation factor (0.462 % at 120 Hz) was obtained. As expected, the values of Tg δ decrease with the increase of the frequency for all sintering conditions.

Table 4. Results of dissipation factor (Tg δ) for the alloy Nb46wt.%Ta capacitor sintered at different sintering temperatures and times

Sintering		Tg δ (%)		
T (ºC)	T (min)	120 Hz	1 kHz	10 kHz
1450	30	0.963	2.665	4.890
1450	60	0.654	1.910	-
1500	30	0.583	1.695	2.410
1500	60	0.550	-	-
1550	30	0.507	1.780	-
1550	60	0.462	1.482	-

Table 5. Results of equivalent series resistance for the alloy Nb46wt.%Ta capacitor sintered at different sintering temperatures and times

Sintering		Equivalent Series Resistance (Ω)		
T (ºC)	T (min)	120 Hz	1 kHz	10 kHz
1450	30	87	37	30
1450	60	84	40	34
1500	30	66	61	48
1500	60	52	48	45
1550	30	33	29	26
1550	60	14	9	6

The values of equivalent series resistance (ESR) for the alloy Nb46wt.%Ta capacitor obtained under different sintering conditions are given in Table 5. This electrical property should be as low as possible. The resistive effects represented by the ESR can affect the capacitor performance in several ways [25]. In general, a decrease can be observed in the ESR values as the sintering temperature is raised. Also, the values of ESR decrease with the frequency.

Table 6 shows the values of leakage current for the alloy Nb46wt.%Ta capacitor obtained under different sintering conditions and bias voltages. It was found that the leakage current decreased for higher sintering temperatures and times. This signify that the quality of the anodic oxide films grown on the alloy Nb46wt.%Ta are strongly influenced by the sintering conditions. These results are in accordance with the values of Tg δ and ESR. As expected, the values of leakage current increase with the bias voltage.

Table 6. Results of leakage current for the alloy Nb46wt.%Ta capacitor sintered at different sintering temperatures and times

Sintering	Leakage Current (μA)				
(°C / min)	2V	5V	10V	15V	20V
1450 / 30	1.2	5.0	16.0	28.0	82.0
1450 / 60	0.9	4.0	11.0	-	25.0
1500 / 30	0.8	1.0	5.0	-	-
1500 / 60	0.8	1.0	5.0	20.0	-
1550 / 30	0.7	1.0	5.0	12.0	6.0
1550 / 60	0.5	0.8	3.4	-	-

Table 7. Results of breakdown voltage for the alloy Nb46wt.%Ta capacitor sintered at different sintering temperatures and times

Sintering		Breakdown Voltage (V)
T (°C)	T (min)	
1450	30	125
1450	60	150
1500	30	160
1500	60	170
1550	30	180
1550	60	-

The values of breakdown voltage (BDV) for the alloy Nb46wt.%Ta capacitor obtained under different sintering conditions are given in Table 7. This property is very important, since this determines the maximum formation voltage with which the sintered capacitor anodes can be charged. The value of BDV increases if the sintering temperature and the sintering time rise. At 1550 °C for 30 min the higher BDV was obtained (180 V). The reason for this is the purification taking place on the anodic surface by the impurities volatization. In addition, the pore structure also contributes for higher values of BDV.

Table 8 shows the values of the specific charge for the alloy Nb46wt.%Ta capacitor obtained under different sintering conditions. The specific charge (CV/g) is defined as the product of the specific capacitance (C/g) times anodizing voltage (V). This property enables monitoring of changes in the specific surface, as well as the capacitance of sintered anodes used in electrolytic capacitors. The data in Table 8 showed that the specific charge of the alloy Nb46wt.%Ta electrolytic capacitor varies strongly with the sintering conditions and frequency used. This signify that the alloy Nb46wt.%Ta powder presents high sintering sensitivity. For higher sintering temperature (1550 °C) the grain growth taking place with a corresponding surface loss.

Table 8. Results of specific charge (CV/g)) for the alloy Nb46wt.%Ta capacitor sintered at different sintering temperatures and times

Sintering		CV/g (μF.V/g)		
T (°C)	T (min)	120 Hz	1 kHz	10 kHz
1450	30	11,270	6,160	2,380
1450	60	16,520	5,250	1,610
1500	30	14,000	5,390	700
1500	60	11,410	4,900	490
1550	30	11,200	6,160	2,380
1550	60	10,640	6,440	2,800

The behavior of the specific charge for the alloy Nb46wt.%Ta electrolytic capacitor at 120 Hz is shown in Fig. 7. This value of frequency is commonly used for the capacitance measurements of tantalum electrolytic capacitors. It can be observed that the alloy Nb46wt.%Ta electrolytic capacitor present a high relatively specific charge from 10,640 to 16,520 μF.V/g. At 1450 °C for 60 min, however, the higher value of CV/g was obtained. This can be explained, because under this sintering condition the densification of the sintered anode is very small, resulting in higher specific surface area (Fig. 4).

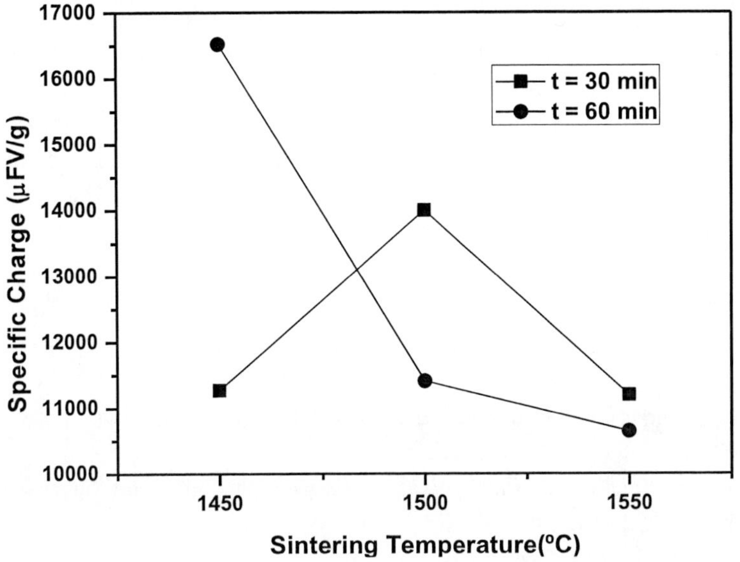

Figure 7. Specific charge of the alloy Nb46wt.%Ta capacitors measured at 120 Hz.

The basic raw material for the manufacture of electrolytic capacitors of high performance is powder, which must have certain quality requirements. The capacitor powder also must be economically competitive. This signifies that per gram of capacitor powder the highest value of CV (amount of electrical charge) possible should be achieved. On the other hand, the high-CV capacitor powders have a series of problems such as higher sintering sensitivity. Thus, a small deterioration in the electrical characteristics (leakage current and breakdown voltage) should be accepted. In this context, it is reasonable to assume that 1450 °C for 60 min and 1500 °C for 30 min are the optimum sintering conditions for the manufacture of sintered anode for the alloy Nb46wt.%Ta electrolytic capacitor.

Conclusion

In this chapter the alloy Nb46wt.%Ta powder for use in electrolytic capacitors was investigated. The results showed that the alloy Nb46wt.%Ta powder display good physical and electrical properties on a laboratory scale for the manufacture of porous anodes for electrolytic capacitors. This powder present high value of specific charge from 10,640 to 16,520 μF.V/g. In addition, the alloy Nb46wt.%Ta powder also present a low cost of production. The sintering conditions 1450 ºC for 60 min and 1500 ºC for 30 min were the more appropriate to yield higher capacitance values per gram with only slight deterioration of the electrical properties (leakage current and breakdown voltage). Therefore, such alloy Nb46wt.%Ta powder produced in fine granulometry (< 20 μm) is very well suitable for production of porous anodes for electrolytic capacitors of high performance.

REFERENCES

Dorf, R.C. *Introduction to electric circuits*. John Wiley & Sons: New York, 1989.

Halliday, D.; Resnick, R. *Fundaments of Physics*. 3rd ed.; John Wiley & Sons: New York, 1988.

Trotter, D.M. *Scientific Am*. 1988, 259, 86-90.

Kahn, M. Ceramic capacitor technology, Electronic Ceramic – Properties, Devices and Applications. Ed. Lionel Levimon: New York, 1988.

Callister Jr., W.D. *Materials science and engineering – an introduction*. 3rd ed.; John Wiley & Sons: New York, 1994.

Sisco, F.T.; Epremium, E. *Columbium and tantalum: applications of columbium and tantalum*. John Wiley & Sons: New York, 1963.

Bernard, W.J. *J. Electrochem. Soc.: Reviews and News*. 1977, 124, 403-409.

Adachi, H. *TIC*, 1991, 68, 3-6.

Mooy, W.F. *TIC*, 1981, 26, 3-4.

Mudiolyubov, Y.M. *TIC*, 1992, 3-4.

Korinek, G.J. *TIC*, 1994, 80, 3-6.

Schile, E.K. *U.S. Patent 4,540,403*, 1985.

Izumi, T. *U.S. Patent 4,645,533*, 1987.

Hohn, R. *U.S. Patent 4,231,790*, 1980.

Holanda, J.N.F. *Thesis Ph.D*, FAENQUIL-DEMAR, Lorena, SP, 1995.

Franklin, R.W. *Proceed. of the First International Tantalum Conference*, 1978.

Kubo, Y.; Igarashi, H. *U.S. Patent 4331*, 1982.

Belz, L.H. *Ta and Nb in some electronic applications*, ASTM, USA, 1984.

Shimizu, S.; Kubo, Y.; Suzuki, T.; Kizaki, T.; Igarashi, H. *U.S. Patent 468*, 1984.

Gomes, U.U. *Thesis Ph.D*, IFGW-UNICAMP, Campinas, SP, 1984.

Gomes, U.U.; Silva, A.G.P.; Holanda, J.N.F.; Pinatti, D.G. *Int. J. Refract. Met. & Hard Mater.* 1992, 11, 43-47.

HCST, *Met. Powder Rep.* 1997, 32, 486-489.

German, R.M.; Munir, Z.A. *J. Am. Ceram. Soc.* 1976, 56, 379-383.

D′yakanov, M.N.; Khomylev, A.F.; Gracheva, G.M. *Sov. Powder Mettal. Ceram.* 1977, 16, 911-915.

Millman, W.A. *TIC*, 1992, 69, 1-8.

In: Manufacturing Engineering
Editors: Anthony B. Savarese
ISBN: 978-1-61209-987-3

Chapter 7

RECENT DEVELOPMENT OF AUTOMATED AND HIGH SPEED MANUFACTURING SYSTEMS

Zheng (Jeremy) Li
University of Bridgeport, Connecticut, USA

ABSTRACT

This chapter is to describe the recent development and applications of automated and high speed manufacturing technology in industrial production. It includes the general application of programmable logic control, studies and design of automated and high speed product assembly line, computer aided design of automated manufacturing systems, computer aided manufacturing simulation, and future trend of automated manufacturing technology. Several case studies in this chapter aim at the introduction, study and analysis for automated and high speed manufacturing and production. The application of programmable logic control to industry brings revolution for the manufacturing techniques. It allows more sophisticated, flexible, reliable, and cost-effective manufacturing process control. Automation is to use control system to reduce human labor intervention during manufacturing processes and production. It plays very important role and puts strong impact in today's industries. Automation is not only significantly increasing the production speed but also more accurately controlling product quality. The automated manufacturing can maintain consistent quality, shorten lead time, simplify material handling, optimize work flow, and meet the product demand for flexibility and convertibility in production. Computer aided engineering design can quickly model the automated manufacturing systems and speed design and development cycle. Computer aided manufacturing can improve engineering integral processes of product design, development, engineering analysis, and production. The current economic globalization requires significant labor cost reduction through industry automation, improved machine tools, and efficient production process.

Based on author's current research projects, several case studies in fully automated and high speed manufacturing systems have been introduced and analyzed in this chapter including fully automated and high speed assembling processes of gas charging system, high viscous liquid filling system, and ultrasonic welding of cap system. All these fully automated and high speed manufacturing systems developed by author have been analyzed and verified through preliminary prototypes or field tests. The computer aided simulation and testing results indicated the reliable performance, feasible function, cost-effective mechanism, increased productivity, and improved product quality by these fully automated and high speed manufacturing systems.

Keywords: automated manufacturing, high speed assembly, computer aided control, process optimization

INTRODUCTION

The development of programmable logic control (PLC) has brought the revolution in manufacturing industry and it provides cost effective, reliable, flexible capability in controlling of complicated manufacturing systems [1], [2]. PLC can be programmed through the technique of relay logic wiring methodology that eliminates the complex of the programming. Automation is applying the control systems and information technologies to reduce the human intervention in manufacturing and production with higher speed and better accuracy. The major advantages of automation include economic enterprise improvement, replacement of human in heavy physical and dangerous working environment [3], [4]. The major advantages of automated manufacturing and production include higher quality, better consistency, reduced production lead times, simplified manufacturing processes, less product handling, and improved work flow in production lines.

Automated manufacturing is a field of engineering dealing with many different manufacturing processes, machine tool research and development, integration of machinery system and production equipments. Robotics applies the mechatronics and automation to manufacturing processes to perform the dangerous, heavy labor related and repetitive tasks [5]. The computer aided engineering (CAE) technology allows the mechanical design teams quickly and cost-effectively iterate the design process to make the product with better quality, cost effective, and reliable function. Through CAE technique, multiple design concepts can be reviewed and evaluated with no real prototype required until the product design completed [6]. Computer aided manufacturing (CAM) controls the production process through computer integrated technology that allows the manufacturing processes exchanging information with each other. The following sections introduce three automated and high speed manufacturing assembly systems from author's recent researches.

1. Automated and High Speed Gas Charging System

The challenge in this automated system design and development is to quickly and reliably seal the high pressure chamber with plug while charging the gas. The normal sealing technologies include injecting the gel into the cartridge for sealing purpose and this system is shown in figure 1. The gel sealing unit is normally used for liquid injection and it shows the poor sealing capacity in high gas charging process.

Figure 1. Gel Sealing System.

1.1. New System Description

In new gas charging and plug sealing mechanism, plug needs to be inserted into cartridge upon gas charging to air chamber since gas will leak out if cartridge is not seal simultaneously. Because of this, gas charging process should be performed in confined space and plug must be assembled into the cartridge precisely with proper tolerance fit.

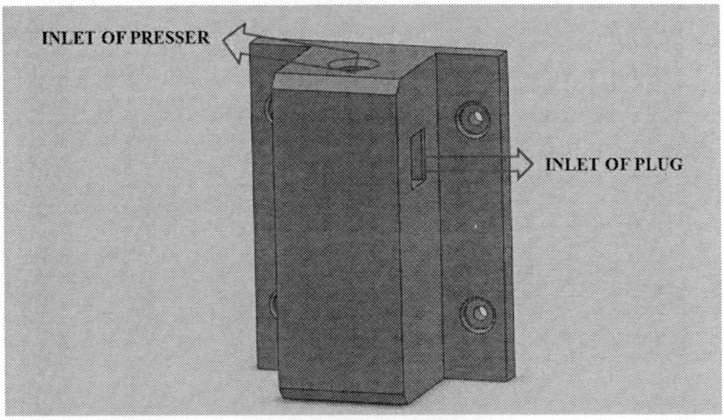

Figure 2. Plug Assembly and Gas Charging Prototype.

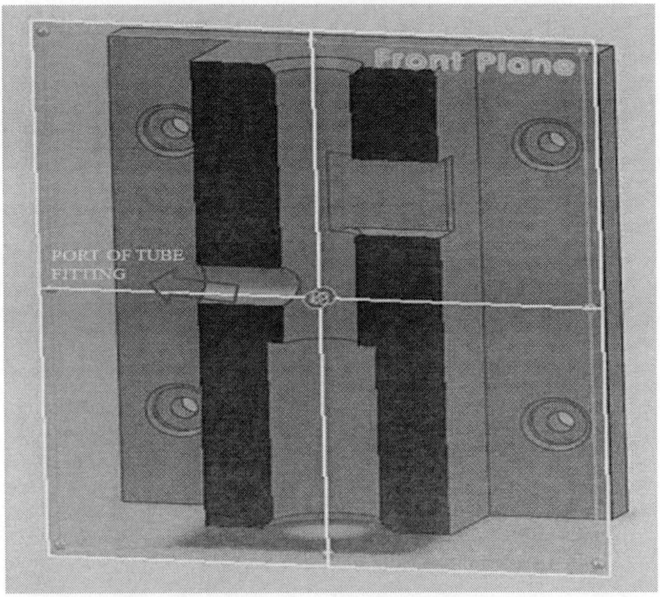

Figure 3. Cross-section View of Plug Assembly Prototype.

The plug assembly system includes plug feeding inlet, central hole functioning as a plug assembly channel, and connecting tube fittings. When empty cartridge automatically delivered to the assembly station, the linear actuator with plug assembly unit attached drives the entire assembly system downward until it contacts the cartridge top surface. Meanwhile, multiple plugs are automatically delivered to the unit by feeding mechanism. The first plug is pushed through the inlet of air chamber to the central hole of instrument. Sequentially, the pusher mounted on the top of center hole is activated to push plug downward with 0.1 inch. The high pressure gas is then injected into sealed chamber. Finally, the top pusher continues to drive plug downward into cartridge for full seal. The detail process is presented in figure 4.

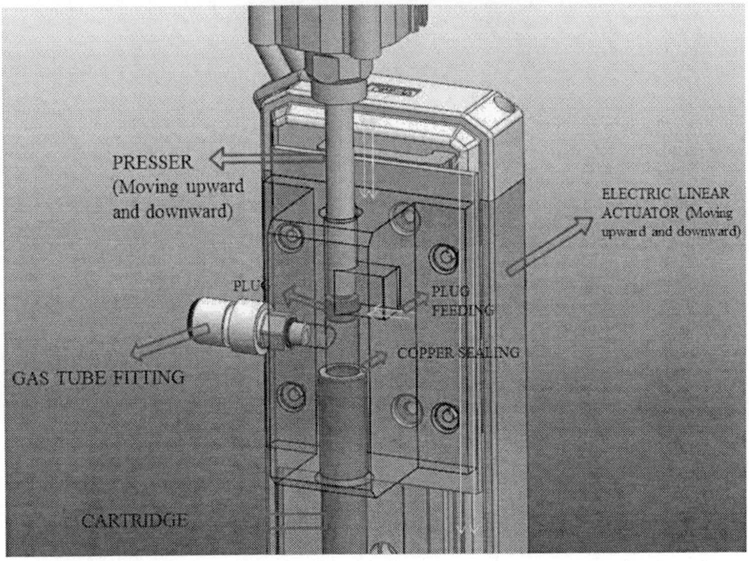

Figure 4. Plug Assembly and Gas Charge Process.

To prevent gas leakage between top surface of cartridge and bottom surface of plug assembly unit, a bronze ring seal is installed, shown in figure 4. This metal ring is pressed and tightly fit at the bottom of center hole in plug assembly unit.

Because the friction force between plug and center hole is larger than the force of high pressure gas, there is no gas leakage from top area of center hole.

If F is gas pressure force,

$$P = \frac{F}{S} \rightarrow F = P \times S = P \times \pi \times \left(\frac{D}{2}\right)^2 = 100psi \times \pi \times (\frac{0.3in}{2})^2$$
$$= 7.0686lbf$$

$P - Maximum\ High\ Air\ Pressure$

$S - Area\ of\ Plug\ Bottom\ Surface$

$D - Diameter\ of\ Plug$

Figure 5. Bronze Ring Seal.

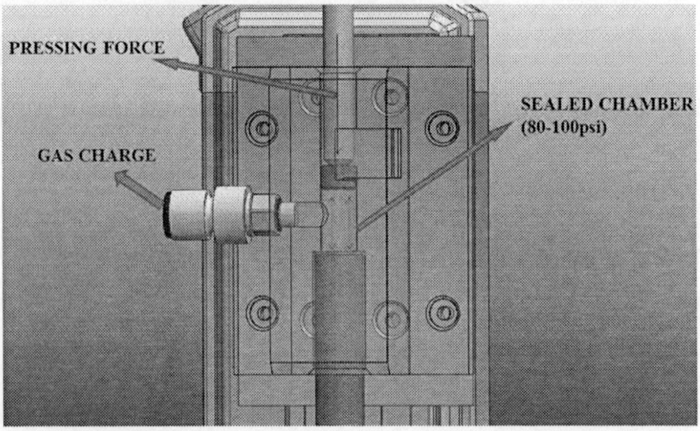

Figure 6. Sealing Chamber.

Table 1 shows the spec of actuator.

Since the pusher is mounted in this actuator, the pushing force 13 lbf is larger than the gas pressure force 7.08 lbf. Therefore, there is no gas leakage in the gas charging process. All the above assembly processes are automated and PLC controlled.

Table 1. Spec of Actuator

Size*	Screw lead	Pushing force [lbf]		Max. speed [mm/s]	Stroke [mm]
		Step motor	Servo motor		
16	10	8.5	6.7	500	50 to 300
	5	16.6	13.0	250	
	2.5	31.7	25.0	125	
25	12	27.4	7.9	500	50 to 400
	6	53.5	16.2	250	
	3	101.6	29.2	125	
32	16	42.5	—	500	50 to 500
	8	83.2		250	
	4	158.9		125	

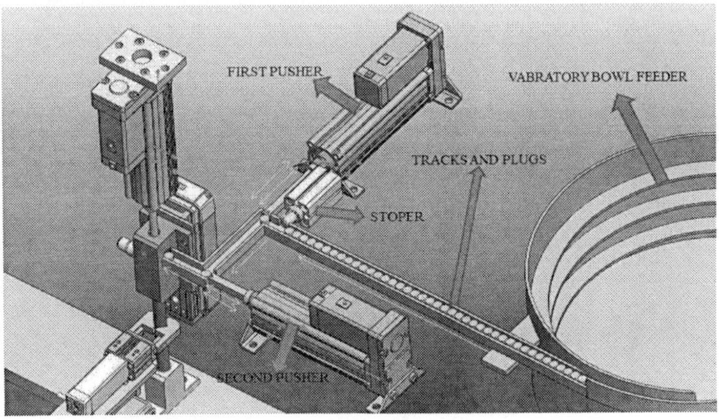

Figure 7. Automated Plug Delivery System.

Figure 10 shows the close view on plug assembly and feeding processes. The plug is fed into the central channel before it is pushed into assembly unit. Top pusher needs to drive plug a little deep which can secure the component assembly.

Due to limited mounting space for plug stopper, a tip portion is designed to compensate the height difference between plug stopper and delivery track system.

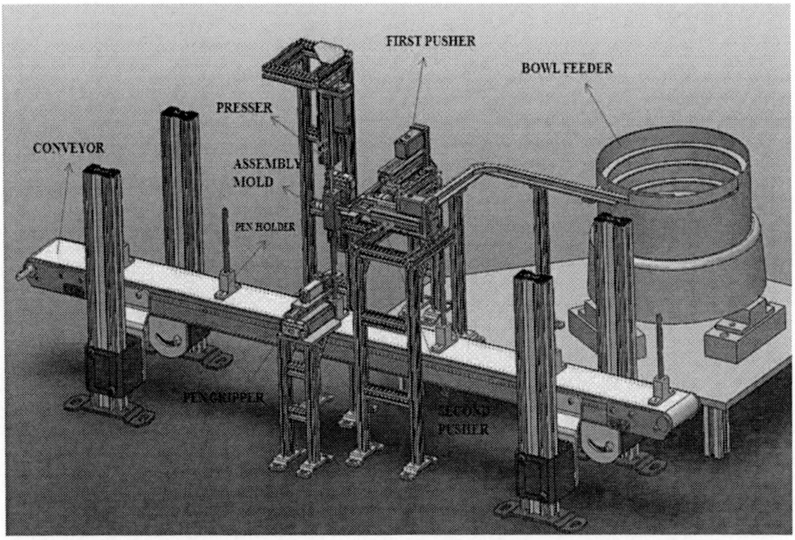

Figure 8. View of Full System Layout.

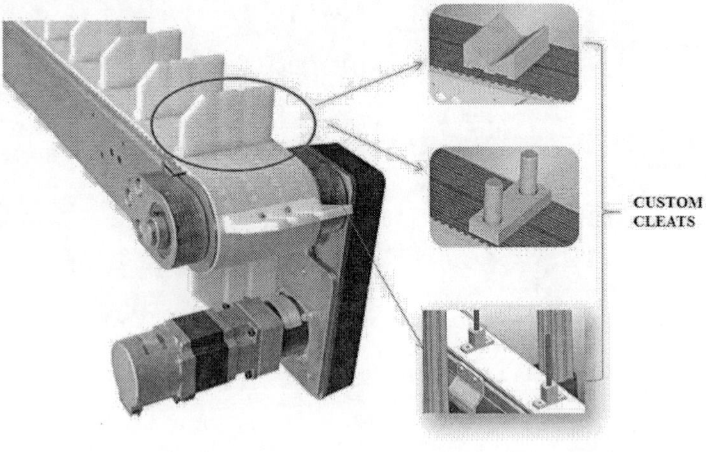

Figure 9. Indexing Conveyor.

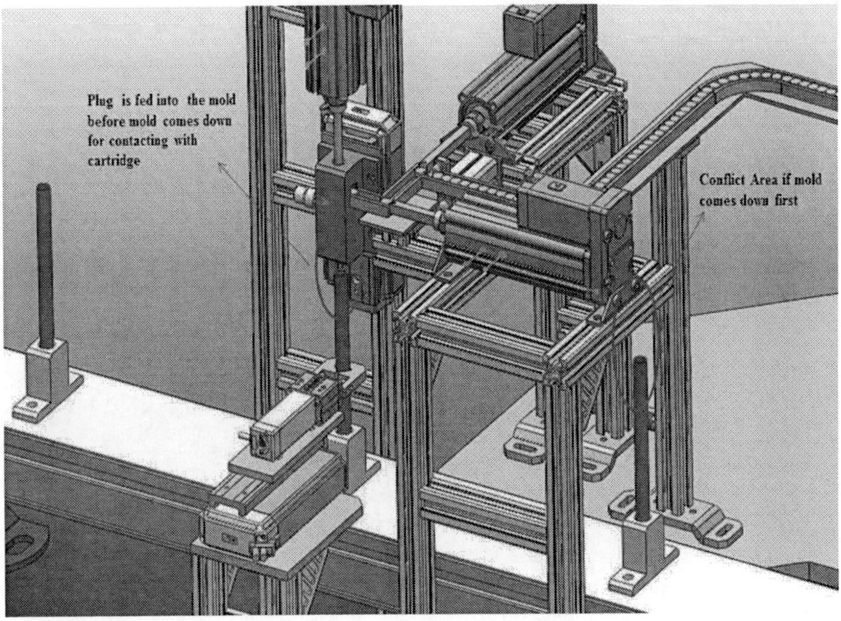

Figure 10. close view on plug assembly and feeding processes.

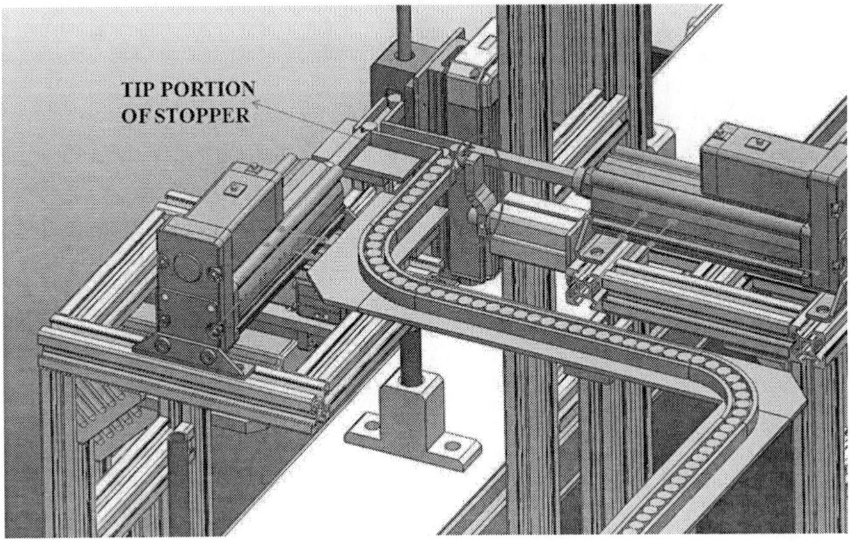

Figure 11. Rear View of Plug Delivery System.

Following two figures give more detail views in different viewing angles.

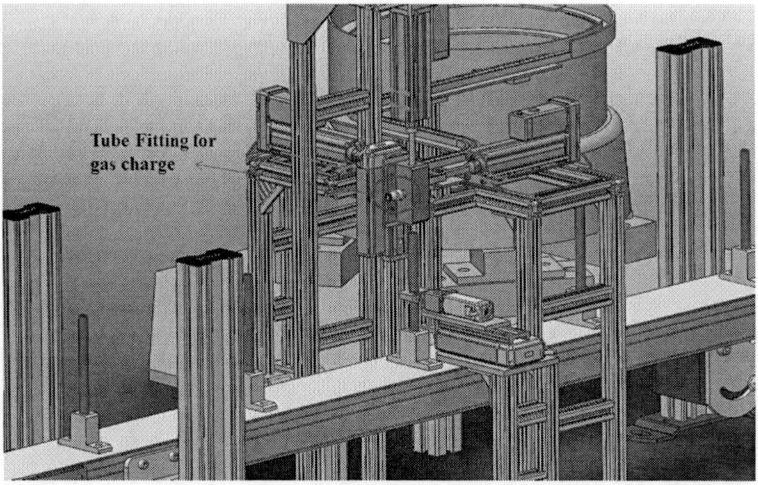

Figure 12. Partial View on Gas Charging System.

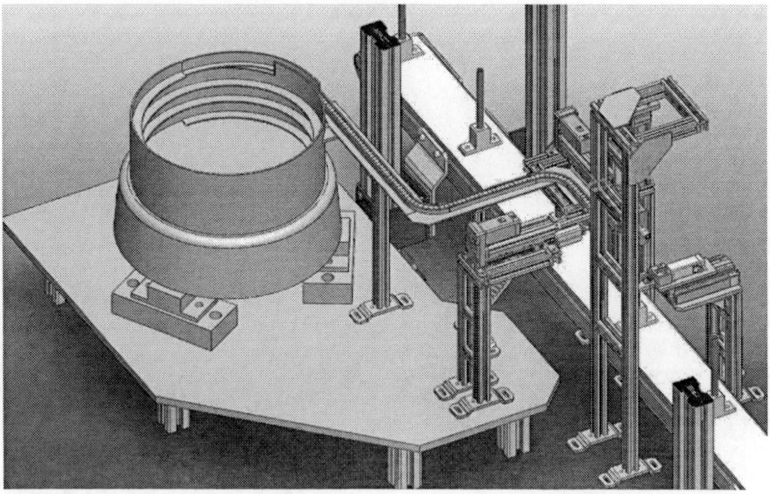

Figure 13. View on Plug Delivery System.

1.2. Stress & Strain Analysis on Cartridge

From above calculation, the force of 7.07 lbf is required to insert the plug into cartridge. COSMOS is used for finite element analysis (FEA) to determine the strength/stress and strain in cartridge during insertion process. The following table shows the material properties of cartridge.

Table.2. Properties of Cartridge Material

Material name:	PS Medium/High Flow
Description:	
Material Source:	
Material Model Type:	Linear Elastic Isotropic
Default Failure Criterion:	Max von Mises Stress
Application Data:	

Property Name	Value	Units
Elastic modulus	2.28e+009	N/m^2
Poisson's ratio	0.387	NA
Shear modulus	8.173e+008	N/m^2
Mass density	1040	kg/m^3
Tensile strength	3.59e+007	N/m^2
Thermal conductivity	0.121	W/(m.K)
Specific heat	1691	J/(kg.K)

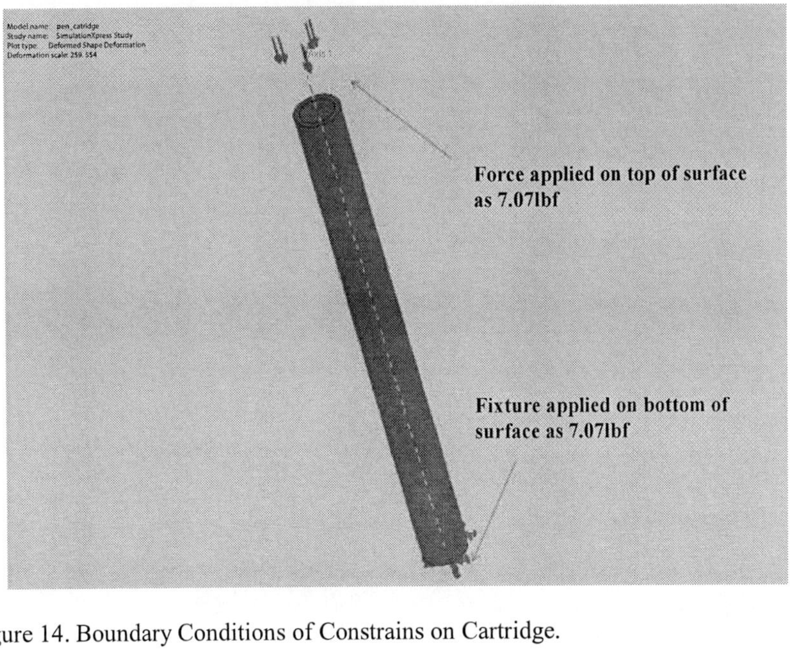

Figure 14. Boundary Conditions of Constrains on Cartridge.

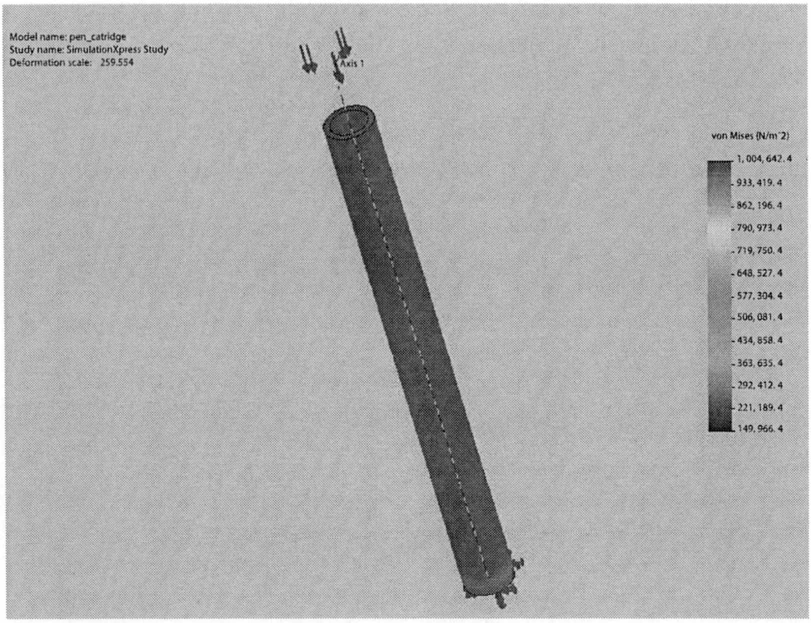

Figure 15. Stress Distribution Inside of Cartridge.

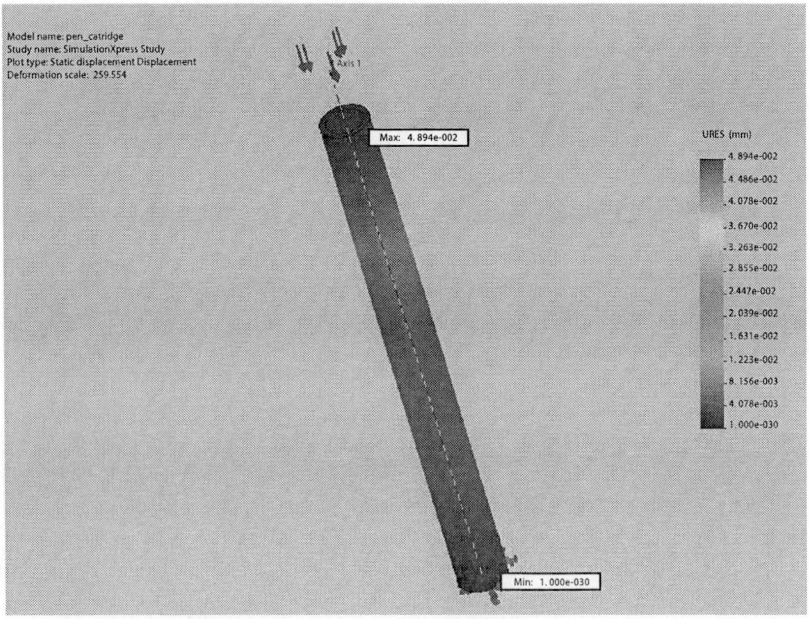

Figure 16. Displacement Inside of Cartridge.

In this stress and strength analysis, 7.07 lbf force is added on the top surface of cartridge and constrain boundary condition is defined at bottom surface of cartridge. The figures 15 and 16 indicate that the maximum stress is 145.70 psi which is lower than material allowable stress, and maximum displacement is 0.0490 inch which is in material safe region.

1.3. Computer Aided Modeling / Simulation

The computational simulation of gas leakage M is calculated as follows:

$$M = B * (T_b / P_b) * D^{2.5} * E * [(P_1^2 - P_2^2) / (L * g * T_a * Z_a * F)]^{0.5}$$

Here,

B - constant

D - pipe diameter

E - pipe efficiency

F - Darcy-Weisbach friction factor

g - gas specific gravity

L - pipe length

P_b - pressure base

P_1 - inlet pressure

P_2 - outlet pressure

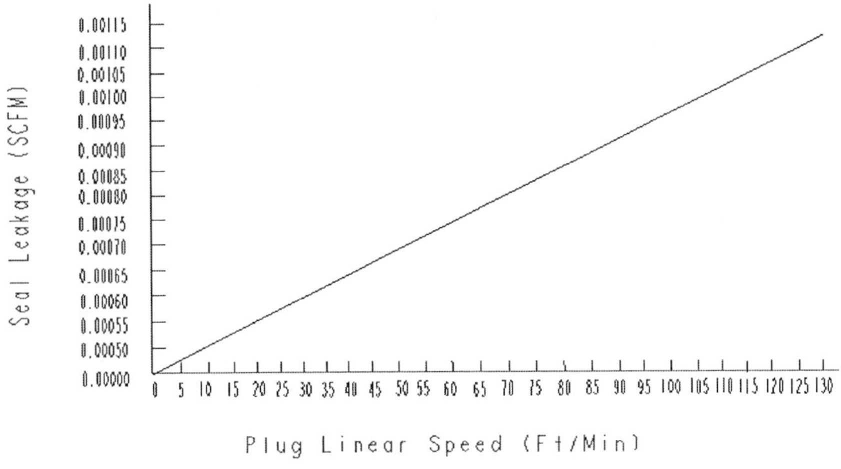

Figure 17. Gas leakage vs. plug linear speed.

M - gas flow rate

T_a - average temperature

T_b - temperature base

Z_a - compressibility factor

The computational simulation of gaseous leakage vs. plug insertion speed is shown in Fig.17.

The prototype of this automated and high speed gas charge system has been tested and gaseous media leakage vs. different plug linear speed is shown in table 1.

Table 1. Gaseous leakage vs. plug linear speed

Piston Linear Speed (Ft/Min)	Estimated Gas Leakage (SCFM)
10	0.0005
20	0.0007
30	0.00085
40	0.00091
50	0.00093
60	0.00095
70	0.00096
80	0.00097
90	0.00099
100	0.00105
110	0.00107
120	0.00110

Both prototype testing and computational simulation show the closed results with good sealing function that verifies the creditability and feasibility of this new automated and high speed gaseous charging system

1.4. Broader Impact

Because of its safe, durable, and reliable functionality, this automated and high speed gaseous charging technology can be applied to many other industrial applications including chemical, natural gas, pharmaceutical, cosmetic, aerospace, defense, military, and biomedical industries. More multiple functioning, sophisticated and cost-effective systems could be designed and developed based on this research project.

2. Automated and High Speed High Viscous Liquid Filling System

The following figure shows the prototype of automated and high speed filling and injection assembly line for high viscous liquids.

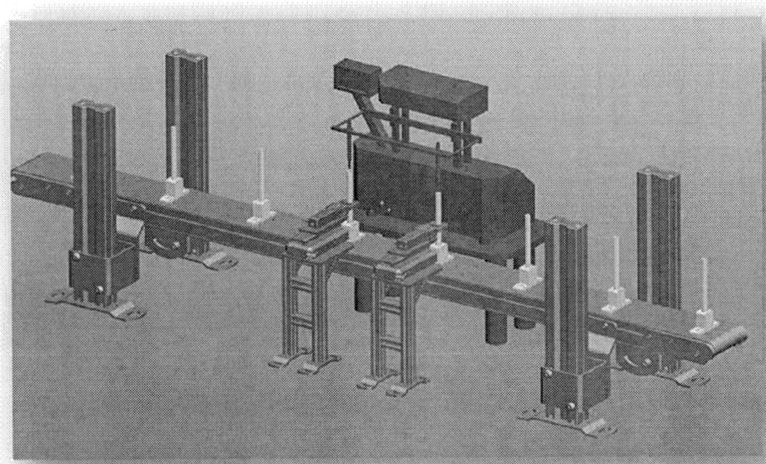

Figure 18 Prototype of Automated and High Speed Filling and Injection Assembly Line for High Viscous Liquids.

The master computer system in this prototype can independently tracks each pump head to determine precisely how much liquid media has been delivered. When the target fill volume is reached, each pump and nozzle is instantly shut off; resulting in high accuracy fills of high viscous liquid products. The computer stores all filling parameters in memory for fast changeovers.

2.1. Technological Merit

This automated and high speed filling system is extremely flexible and designed to fill high viscous liquids. It can be applied to many industries including pharmaceutical, cosmetic, dairy, chemical, food, etc. For example, the thin and thick liquids, cosmetic creams, and chunky sauces with pasteurized temperatures can all be filled in this prototyped new system.

In this special filling production, the positive displacement pump will be required to accord with the high viscosity and high temperature liquid media. The rotary gear pump is used in this filling system for heavy duty work including all types of oils, roofing tar, bitumen, ink and wax, etc. Rotary gear pump with jacketed design can work at temperature up to 120℃. It has double driving rotor which makes it to be very powerful for high viscosity liquids. To improve the filling speed, multiple filling nozzles, such as 4, 8, or 12 nozzles, can be designed and installed in inline filling system. The swaged internal nozzle has been used in this system and it is especially applied for narrow bottlenecks and high temperature environment.

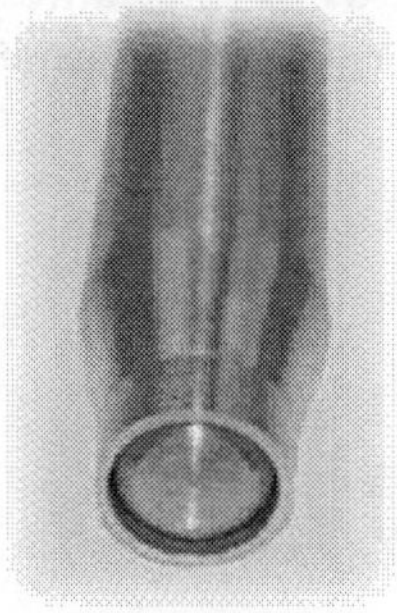

Figure 19 Swaged Internal Nozzle.

2.2. New System Description

The full system of automated high viscous liquid filling system consists of a conveyor system, gripping system and filling mechanism, shown in figure 20. The precision indexing conveyor is used to control the precision indexing tolerance to register a part within tolerance range of 0.015 – 0.20 inch without installation of detection sensors.

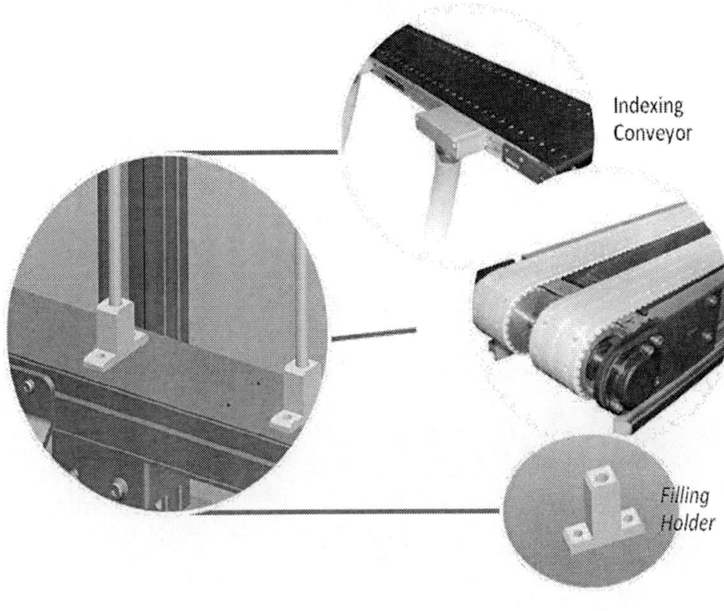

Figure 20. Indexing Conveyor for Filling System.

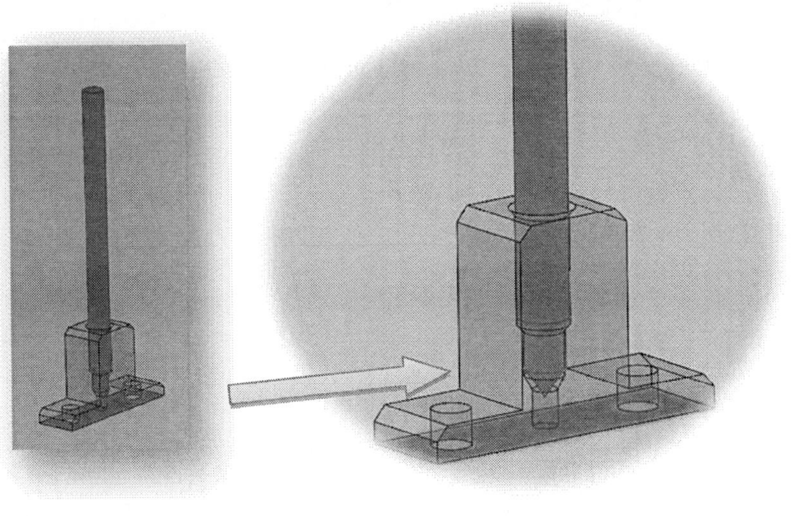

Figure 21. Filling Holder.

The filling holder is mounted on indexing conveyor and the detail of filling holder is shown in figure 21. This filling holder has proper tolerance with outside diameter of cartridge to secure high speed filling process.

The high viscous liquid filling process is described as follows.

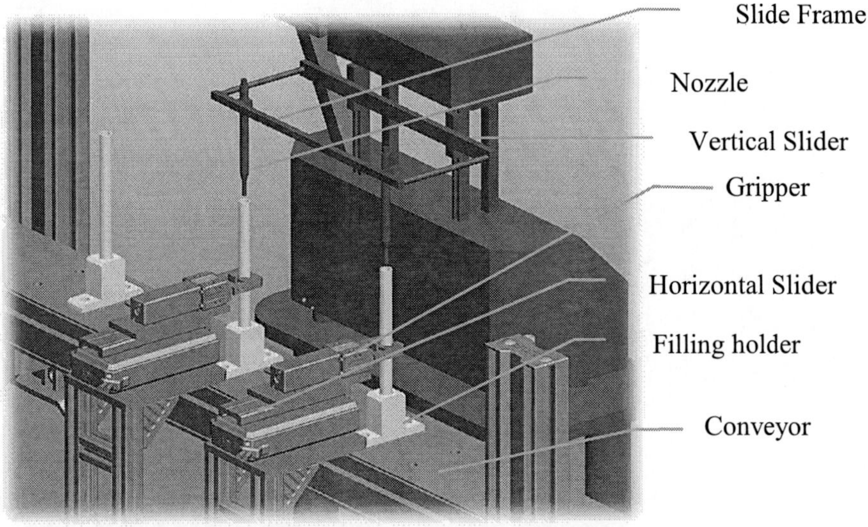

Figure 22. High Viscous Liquid Filling and Injection System.

During production cycles, the cartridge which is inserted in filling holder, moves along with indexing conveyer. When empty cartridges move to the location below filling nozzles, conveyor will be stopped by sequence control of PLC system which receives the holder location signal from opposed mode sensor. Horizontal sliders mounted at the frame of gripper system push gripper pair forward to securely hold cartridges for filling process. The slide frame will then take both nozzles downward by vertical sliders and nozzles are inserted into cartridges to begin filling process. After high viscous liquid filling process, the vertical slide frame moves nozzles upward, and grippers release the cartridges and move away from filling holder allowing more clearance for cartridge moving.

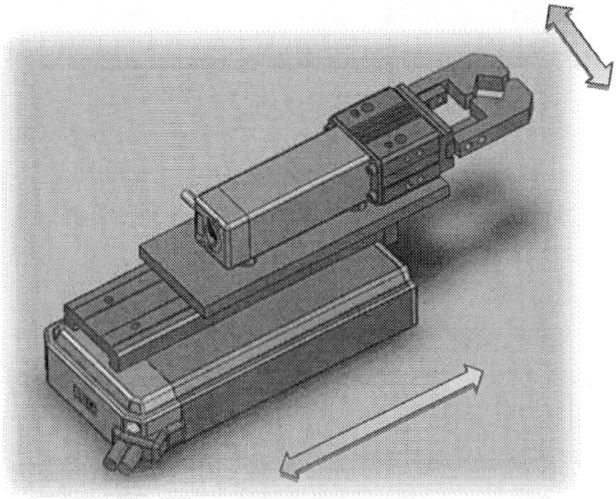

Figure 23. Gripper.

3. Automated and High Speed Cap Sealing

Automated and high speed plastic welding is a high volume manufacturing process to weld plastic components together. It is one of the primary processes of joining or welding plastics. There are several types of techniques in plastic welding processes including hot gas welding, speed tip welding, extrusion welding, contact welding, high frequency welding, hot plate welding, injection welding, ultrasonic welding, friction welding, laser welding and solvent welding. One basic method is using outside hot resource, such as hot air or hot metal tip, to heat up the joining parts for sealing. The high temperature softens the plastic parts and makes them sticky to join multiple surfaces together. The applications of this welding technique include hot gas welding, speeding tip welding, extrusion welding, contact welding and hot plate welding. Another type of welding uses special technology to join components together, such as high frequency welding process through high frequency electromagnetic. The laser welding technique puts work pieces under pressure while laser beam moves along the joining line, and the solvent welding technique uses a dissolve method to join parts together. The friction welding process makes use of vibration among joining surfaces with specified frequencies and amplitudes.

3.1. Technological Merit

The ultrasonic welding is applied in this automated and high speed cap sealing system that involves the vibration from high frequency sound energy to soften or melt the thermoplastic components. The ultrasonic vibration nornally has its frequencies of 20 – 40kHz. When ultrasonic vibrations stop, molten material solidifies and weld is achieved.

The ultrasonic welding is the quickest welding method with welding processing time well under 1 second which is good for high-speed assembly line. The ultrasonic welding in this project not only has the fastest welding speed, but also avoids the damage from traditional cartridge cap sealing method. The traditional cap sealing methods use the mechanical tolerance principle to make tight clearance between two mating parts. The process of assemble using traditional method has to punch the cap into cartridge and it can bring a large swing vibration which could affect the cap location in cartridge.

3.2. New System Description

The cap sealing assembly line includes conveyor system, gripping system, placing system, ultrasonic welding machine and feeding system. The full assembly line is shown in figure 24.

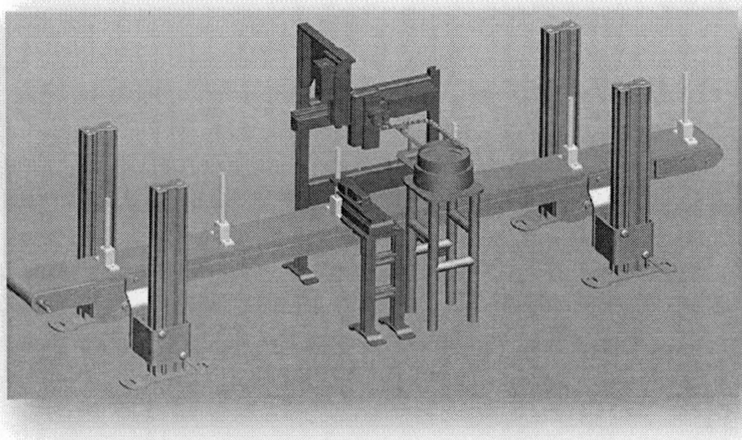

Figure 24. Automated and High Speed Cap Sealing Assembly Line.

The automated and high speed cap welding operation is illustrated as follows:

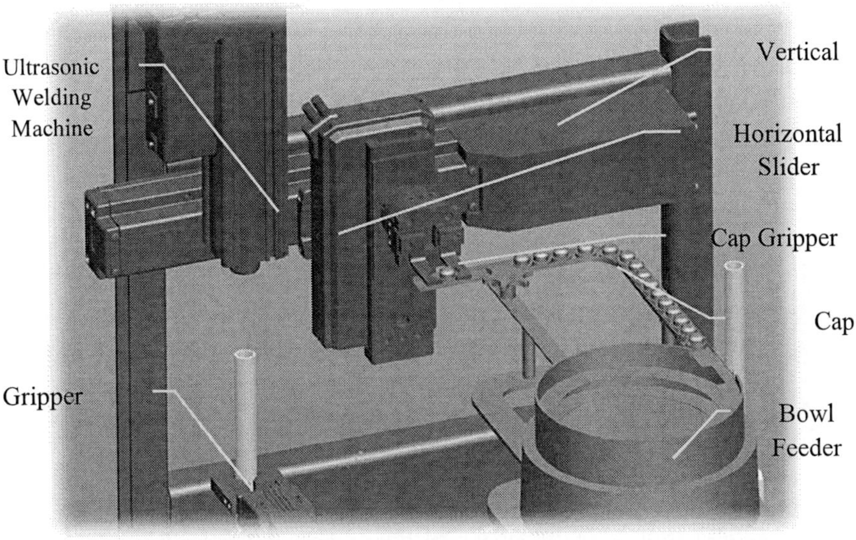

Figure 25. Detail View of Automated and High Speed Cap Welding Operation.

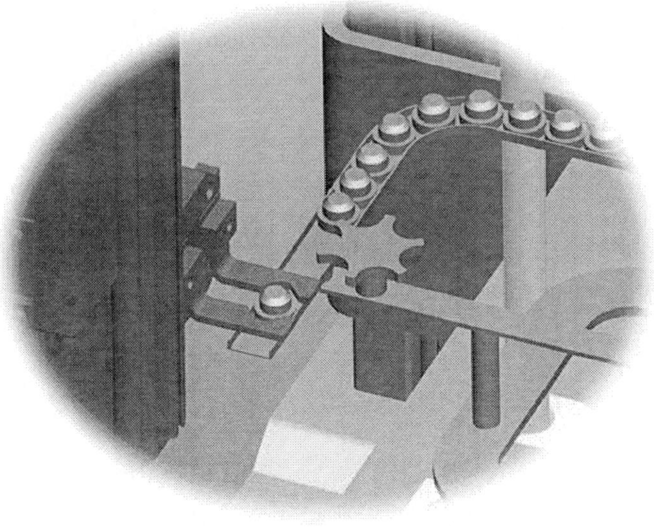

Figure 26. Cap Feeding System.

While cartridge stops at location below the welding machine horn, the gripper moves forward to clamp cartridge for its secured position. The gripper picks up a cap from feeding rail and moves upward by vertical air slider. Then the horizontal slider moves cap gripper and vertical slider drives welding mechanism forward. At the same time, motor brings cap gripper and rotates in 180 degree to make cap bottom upwards. When cap is brought to the cartridge top surface position, horizontal slider in gripping system will move gripper to insert cap into cartridge. The gripper will release and move back to pick up next cap in feeding rail. The ultrasonic welding mechanism moves down by vertical slider and contact the cap for ultrasonic welding.

Figure 26 shows the caps feeding system. The caps will be separated while move forward and adjusted by a rotate wheel which can control rotate degree and speed.

Conclusion

The application of automated and high speed manufacturing technology brings revolution for industrial production with sophisticated, flexible, reliable, and cost-effective manufacturing process control. Automated and high speed manufacturing is to use computer aided control system to reduce human labor and prevent work force from dangerous and hazard environment in production. The automated and high speed manufacturing technology plays critical role and adds significant impact in modern industries. This technology is not only increasing the production quantity but also improving manufacturing quality. The several research projects of automated and high speed manufacturing systems developed by author can help to keep consistent product quality, reduce production lead time, speed up material handling, improve work flow, and satisfy customer on their product demand with flexibility and convertibility in manufacturing process and mass production.

References

Kundu, Pijush K., Cohen, Ira M. (2008), Fluid Mechanics (4th revised ed.), Academic Press, ISBN 978-0-123-73735-9

Isaev S.A., Baranov P.A., Kudryavtsev N.A., Lysenko D.A., Usachov A.E. (2005) Comparative analysis of the calculation data on an unsteady flow

around a circular cylinder obtained using the VP2/3 and Fluent packages and the Spalart-Allmaras and Menter turbulence models J.
Engineering Physics and Thermophysics.Vol.78.No.6. pp. 1199-2013.

Norberg, C. (2003). Fluctuating lift on a circular cylinder: Review and new measurements, Journal of Fluids and Structures 17 (1), pp. 57-96 29

Zdravkovich, M.M. (2003), Flow Around Circular Cylinders, Vol. 2 Applications, Oxford University Press.

Chakraborty, J., Verma, N. and Chhabra, R. P. (2004), Wall effects in the flow past a circular cylinder in a plane channel: a numerical study, Chem. Engng. Processing, 43: 1529-1537.

White, Frank M. (2003), Fluid Mechanics, McGraw–Hill, ISBN 0072402172

INDEX

C

D

G

H

I

J

K

L

M

N

O

P

Q

R

S

T

U

V

W

X

Y

Z